Petroleum Engineering Handbook

Petroleum Engineering Handbook

Joanna Carter

MURPHY & MOORE
www.murphy-moorepublishing.com

Petroleum Engineering Handbook
Joanna Carter
ISBN: 978-1-63987-426-2 (Hardback)

MURPHY & MOORE
Published by Murphy & Moore Publishing,
1 Rockefeller Plaza,
New York City, NY 10020, USA

Cataloging-in-Publication Data

Petroleum engineering handbook / Joanna Carter.
p. cm.
Includes bibliographical references and index.
ISBN 978-1-63987-426-2
1. Petroleum engineering. 2. Petroleum--Prospecting. 3. Mining engineering.
I. Carter, Joanna.
TN870 .P48 2022
665.5--dc23

For more information regarding Murphy & Moore Publishing and its products, please visit the publisher's website www.murphy-moorepublishing.com

Table of Contents

Preface

The field of engineering which is concerned with hydrocarbon production is known as petroleum engineering. It is a multi-disciplinary field that applies the principles of mechanical engineering, chemical engineering, mining engineering and physics. Petroleum engineering is divided into various sub fields such as reservoir engineering, drilling engineering and petrophysics. Its key objective is to extract liquid and gaseous hydrocarbons from the Earth's surface. This requires estimation of recoverable volume and a detailed understanding of water and oil within a porous rock under very high pressures. The processes used in petroleum engineering are divided into upstream, mid-stream and downstream sectors. The upstream activities involve searching for oil and gas fields, which are located underground or underwater. Midstream sector is related to the transportation of oil and gas. The downstream processes focus on refining of crude oil to obtain gasoline. The various sub-fields of petroleum engineering along with technological progress that have future implications are glanced at in this book. The topics covered herein deal with the core subjects of petroleum engineering. This book will serve as a valuable source of reference for those interested in this field.

To facilitate a deeper understanding of the contents of this book a short introduction of every chapter is written below:

Chapter 1- The branch of engineering which deals with the activities involved in the production of crude oils and natural gases is known as petroleum engineering. Some of the topics related to petroleum engineering like petroleum and crude oil are explained in detail. This is an introductory chapter which will introduce briefly all significant aspects of petroleum engineering.

Chapter 2- The search for deposits of petroleum and natural gas in the Earth using petroleum geology is known as petroleum exploration. Petroleum extraction is the process by which petroleum is extracted from the deposits beneath Earth's surface. This chapter discusses in detail the various methods of petroleum extraction like oil well, drilling technique, mud logger, drilling rig and deep well drilling, etc.

Chapter 3- The pool of hydrocarbons present in permeable rocks under the surface of Earth is known as petroleum reservoir. The branch of petroleum engineering which involves applied reservoir mechanics for the development of oil and gas reservoirs is known as reservoir engineering. This chapter has been carefully written to provide an easy understanding of the varied facets of petroleum reservoir and petroleum engineering.

Chapter 4- The method used to extract petroleum or natural gas by fracturing of rocks by a pressurized liquid is known as hydraulic fracturing. This chapter encompasses waterless fracturing, hydraulic fracturing proppants, regulation of hydraulic fracturing, hydro-slotted perforation and uses of radioactivity in oil and gas wells. The aim of this chapter is to explore the field of hydraulic fracturing.

Chapter 5- The recovery of hydrocarbons refers to the extraction of crude oil and natural gases. It involves various steps; primary recovery, secondary recovery, tertiary recovery and infill recovery. Primary and secondary recovery processes accounts for the 20-40% recovery of oils while tertiary recovery can extract 30-60% more reservoir oil. This chapter discusses in detail the theories related to enhanced oil recovery.

Chapter 6- The industrial process plant which transforms crude oil into various useful products like petroleum, gasoline, kerosene, liquefied petroleum gas, etc. This chapter covers amine gas treating, catalytic reforming, cracking, fluid catalytic cracking, gas flare, hydrodenitrogenation and hydrodesulfurization in an easy-to-understand manner. The topics elaborated in this chapter will help in gaining a better perspective about petroleum refinery.

Finally, I would like to thank the entire team involved in the inception of this book for their valuable time and contribution. This book would not have been possible without their efforts. I would also like to thank my friends and family for their constant support.

Joanna Carter

1

Petroleum Engineering: An Introduction

The branch of engineering which deals with the activities involved in the production of crude oils and natural gases is known as petroleum engineering. Some of the topics related to petroleum engineering like petroleum and crude oil are explained in detail. This is an introductory chapter which will introduce briefly all significant aspects of petroleum engineering.

Petroleum

The word 'petroleum' is derived from a Greek word, which means "rock oil". Petroleum is used since ancient times for various purposes. Crude oil is a mixture of several thousands of hydrocarbon compounds of various molecular weights. Although petroleum can form by inorganic processes, organically formed petroleum is much more significant. Petroleum consists almost entirely of hydrogen and carbon, which has the ratio of 1.85 hydrogen atoms to one-carbon atoms in crude oil. Sulphur, nitrogen and oxygen comprises up to 3% of most petroleum. In its strict sense, petroleum includes only crude oil, but it occurs as liquid, gaseous and solid hydrocarbons. Petroleum may occur as solid state in forms of tar sands, natural asphalt, gibsonite, albertite, grahamite, oil sands and oil shales. These are known as non-conventional hydrocarbon resources. The liquid state of petroleum includes the conventional heavy and light crudes and condensates. Condensates occur in the form of gases in the subsurface, but converts to liquid in the surface. Gaseous petroleum includes conventional hydrocarbon gases, methane hydrates, swamp gas, coal bed methane, shale gas.

Table: Different states of petroleum.

State of petroleum	Examples
Solid	Gibsonite, albertite, natural asphalt, tar sand, oil shale.
Liquid petroleum	Conventional light crude oil and heavy crude oil condensate.
Gaseous petroleum	Conventional (natural gas) unconventional (methane hydrate, swamp gas, coal bed methane, shale gas).

Composition of Petroleum

Petroleum is essentially made up of carbon and hydrogen. The other elements that may be present are Nitrogen, Sulphur and Oxygen. A very small amount of metal is occasionally detected in liquid hydrocarbon. Hydrocarbon compounds maybe classified as Paraffins, Naphthenes, Olefins, Aromatics, Naphtheno-aromatics and ashphaltic compounds.

Paraffins

Paraffins are alkanes or straight-chained hydrocarbon. They are saturated hydrocarbon having the general formulae C_nH_{2n+2}. They form a homologous series where the next compound of the series is obtained by adding CH_2 to the previous compound. The members of this series having low n value, i.e. from 1 to 4 are gaseous namely, methane, ethane, propane and butane. The paraffins with n values from 5 to 17 occur as liquid form. As the n value of paraffin increases, its wax content increases and ultimately it converts into a semi-solid form. Branched chain paraffins are known as isoparaffins. Although the chemical formula of the straight chain and branched chain paraffin are the same, physical properties of crude oil, like boiling point vary. Thus the boiling point of normal butane is (-0.5 °C), while in case of isobutane it is -12 °C.

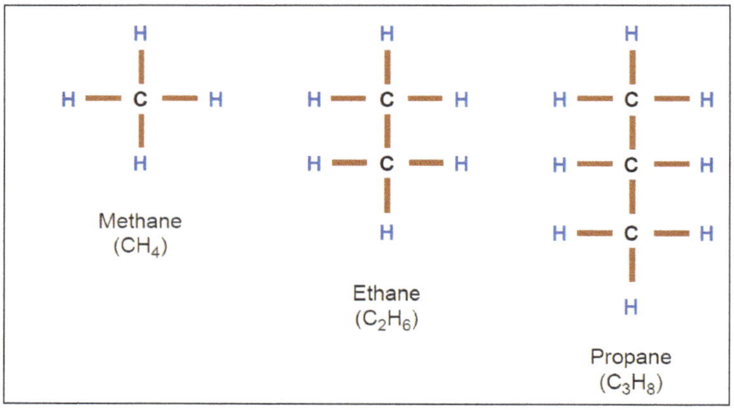

Example of alkanes.

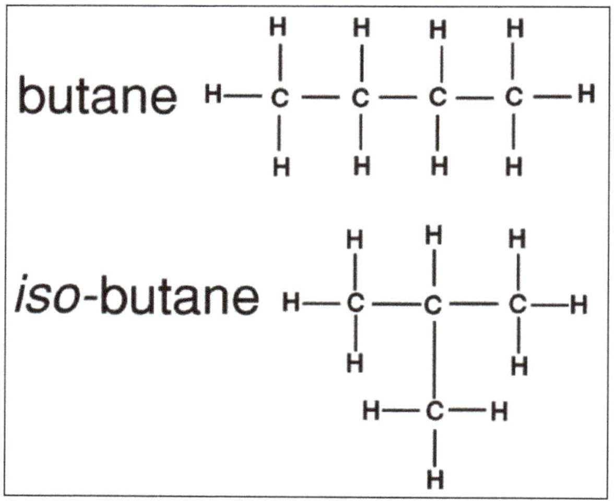

Isomers of butane (n-butane and iso-butane).

Naphthenes

Naphthenes are cyclo-alkanes consisting of saturated hydrocarbons which have one or more carbon rings to which hydrogen atoms are attached according to the formula C_nH_{2n}. Naphthene is the most common hydrocarbon compound of crude oil constituting about 50%. Cycloalkane rings larger than C_7 or smaller than C_5 are not found in crude oil. The hydrocarbon rings may merge together to form complex molecules.

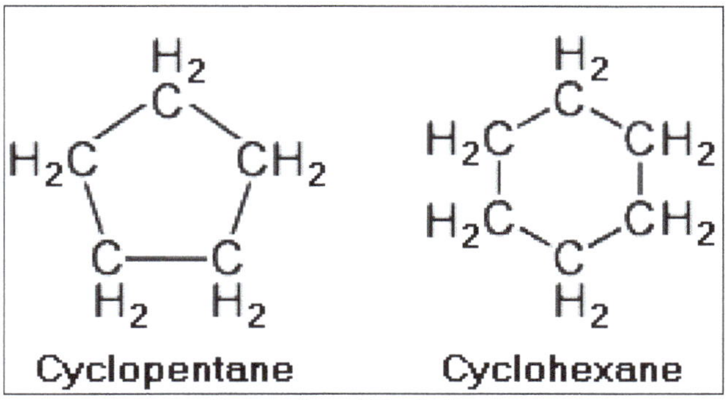

Examples of Naphthene.

Olefins

Olefins are unsaturated alkenes containing double bonds in their structure. They are isomer of cycloalkanes with the general formula C_nH_{2n}. They are very rare in crude oils.

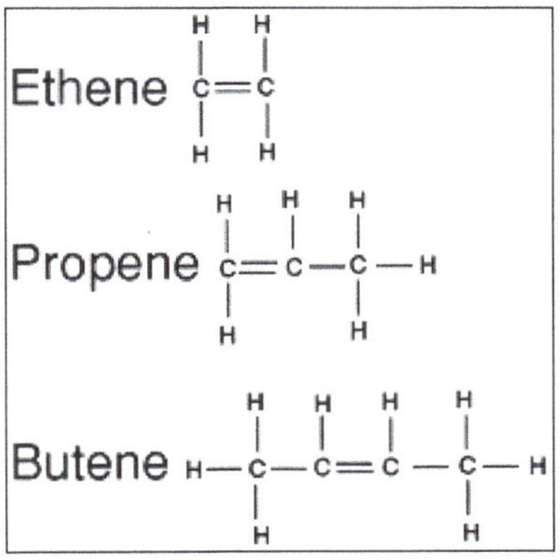

Olefins (ethene, propene, 2 butene, 2, 4 pentadiene).

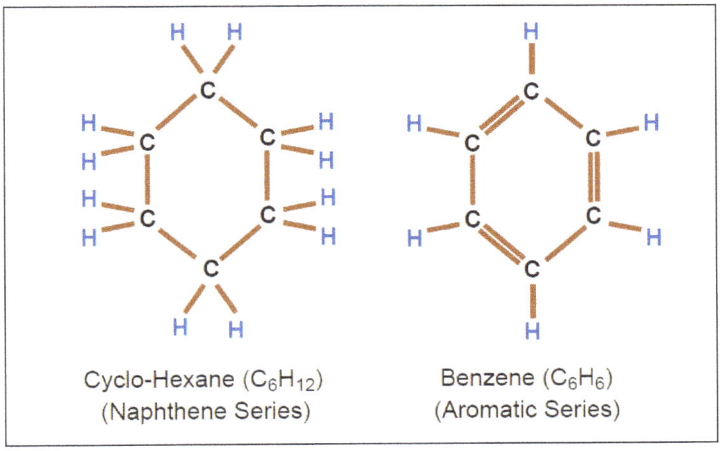

Example of aromatic hydrocarbon and its comparison with naphthene for the same carbon number.

NSO Compounds

NSO compounds include nitrogen, sulphur or oxygen in the molecular structure of the hydrocarbon. They are mainly present in the heavy (large molecule) residual fractions of petroleum.

The most important sulphur compounds are thiols and thiophenes. The free sulphur range in natural crude oils is less than 0.5-5% of petroleum. The nitrogen Compounds that may be present are pyridines, quinolenes, and indoles. The range of nitrogen in natural crude oils is 0.25 to 0.8%. The common oxygen compounds that are present in petroleum are organic acids, alcohol, phenol, esters etc. The range of oxygen in natural crudes is diagrammatically represented in Figure below.

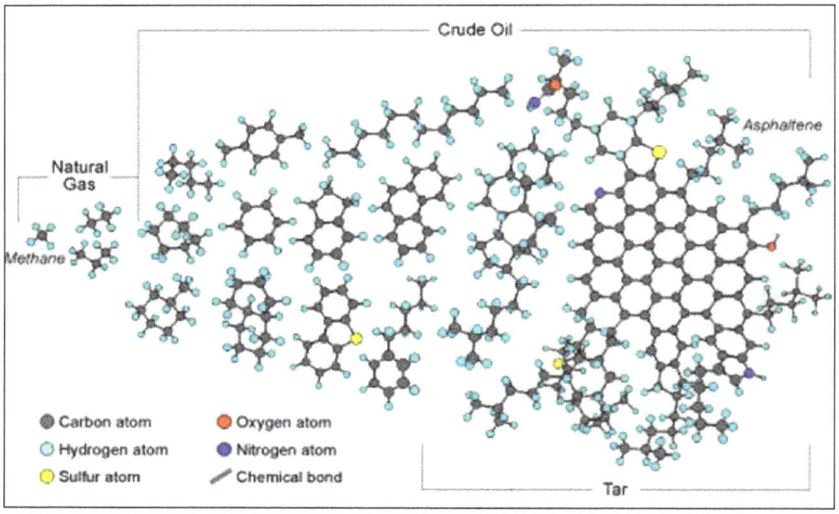

Schematic representation of chemical compounds present in the petroleum.

Crude Oil

Crude oil is a naturally occurring, unrefined petroleum product composed of hydrocarbon deposits and other organic materials. A type of fossil fuel, crude oil can be refined to produce usable products such as gasoline, diesel, and various other forms of petrochemicals. It is a nonrenewable resource, which means that it can't be replaced naturally at the rate we consume it and is, therefore, a limited resource.

Crude oil is typically obtained through drilling, where it is usually found alongside other resources, such as natural gas (which is lighter and therefore sits above the crude oil) and saline water (which is denser and sinks below). It is then refined and processed into a variety of forms, such as gasoline, kerosene, and asphalt, and sold to consumers.

Although it is often called "black gold," crude oil has ranging viscosity and can vary in color from black to yellow depending on its hydrocarbon composition. Distillation, the process by which oil is heated and separated in different components, is the first stage in refining.

Physical and Chemical Properties of Crude Oil

The physical properties of crude oils are the quantitatively measurable characteristics of crude oils. They vary according to the composition of the oil, the relative abundance of the groups of hydrocarbons, and essentially depend on reservoir temperatures and pressures.

Specific (or A.P.I) Gravity

This is the weight of a given volume of crude oil. It is measured in two gravity scales, as stated below:

- $\text{A.P.I. gravity (in degrees)} = \frac{141.5}{\text{SP.gravity at } 60\ ^\circ\text{F}} - 131.5$
- $\text{Baume gravity (in degrees)} = \frac{140}{\text{SP.gravity at } 60\ ^\circ\text{F}} - 130$

The A.P.I. (American Petroleum Institute) gravity scale is more commonly and widely used than the European Baume gravity scale. But they essentially measure the density of a crude oil. The A.P.I. gravity of crude is influenced by the composition of the oil. Crude oils characterized by high amount of dissolved gases, are less dense thus, light in weight and therefore, possess high A.P.I. gravities, while denser crude oils of low amount of dissolved gases are characterized by low A.P.I. gravity values.

The group of hydrocarbons predominating in a crude oil also influences the A.P.I. gravity. For example, paraffin base crude oils (45-60% paraffin hydrocarbons and less amount of naphthenes and aromatics) are light, thus, high A.P.I. gravities. But naphthemic base crudes (consisting predominantly of naphthene hydrocarbons 60-75%, with lesser amount of paraffins and aromatics) are heavy and have low A.P.I gravities? The A.P.l. gravities of crude oils usually increase with depth. This is because a combination of source and reservoir maturation processes, Associated with slow but continuously increasing geotemperatures, cause the generation of lighter (or High A.P.I gravity) oils at greater depths of burial.

Viscosity

This is the measure of resistance to flow in crude oils due to internal friction. It is expressed in 'poise' or 'centipoise'. Crude oil is said to have a viscosity of 1 poise when a tangential force of 1 dyne causes a plane surface of 1 square centimetre area; spaced 1 cm from a stationary plane surface to move with a constant velocity of 1 cm per second, the space being filled with the viscous c.

The viscosity of a crude oil is influenced by the amount of dissolved gases at the prevailing temperature. Crudes characterized by high amount of dissolved gases have high A.P.I gravities and low viscosity or moderately high fluidity. At high temperatures molecular agitation (or velocity) of the crude increases, making for a volumetric expansion and reduction in internal molecular friction, thus, reducing the oil. The greater the quantity of a high-molecular weight hydrocarbon group in a crude, the denser and more viscous.

Refractive Index

The refractive index n, of a crude oil is measured from,

- $n = \frac{\text{Sini}}{\text{Sinr}}$

where; i = incidence angle,

r = angle of refraction.

It depends on the density of the oil. Heavy crudes (of low A.P.I) have high refractive indices. This is because a dense crude would create a dense medium for a passing ray of light, which is refracted towards the normal at a low angle (r). On the other hand, light oils have low refractive indices.

Optical Activity

Optical activity is the power of crude oils to rotate the plane of polarization of a polarized light. It is commonly expressed in degrees per millimetre. If any crude oil causes the plane of polarization to rotate to the right, it is called a 'dextroratory', but if is to the left it is known as a 'levorotary'. This property is destroyed at high temperatures (250 – 300 °C).

Optical activity is also exhibited by some organisms that contain cholesterine substances (such as cholesterol $C_{26}H_{45}OH$). The amount of optical rotation shown by a crude oil depends mainly on its steranepentacyclic and triterpane content. And these are hydrocarbon compounds derived from the microbial decarboxylation of organic cholesterine substances.

Cloud and Pour Points

The pour point is the temperature at which crude oil will no longer flow, when a tube containing it is first heated in a bath, in order to dissolve all its wax content and then gradually cooled. At this temperature, the crude oil is in semi-solid to solid form, and thus loses its fluidity. Slightly above this temperature of no-flow (the pour point), there might had been an appearance of cloudy substances in the crude: which is due to the settling out of the solid paraffin waxes contained in the crude oil. That temperature is called the cloud point. This property determines the influence of low temperatures on crude oils. Simultaneously, it provides information about the amount of solid paraffin waxes contained in the oil. This property is common in paraffin base crudes but wax-free naphthenic oils do not show cloud point. If the pour point of a crude is above the surface temperature, it will precipitate its paraffin waxes on approaching the surface of the ground will only flow on heating. The range of pour points of crude oil varies greatly from -70 °F to +90 °F or higher.

Volume

The volume of a crude oil in its reservoir rock differs from the volume it occupies in the surface. This is due to formation gas-oil ratio and reservoir pressures. The formation gas-oil ratio expresses the volume of gas contained in one barrel of a crude oil as it comes from the reservoir rock. Under high reservoir pressure, the volume of oil in the reservoir increases because of the influence of dissolved gases. But on release of the reservoir pressures, the dissolved gases escape, leading to the shrinkage of the volume of the crude oil at the surface.

Fluorescence

It may be yellow, green or blue. For example, when a paraffin base crude oil (gasoline-rich) is exposed to ultraviolet fluorescence light, it emits yellow colour, while naphthenic oils emit brownish colour This property is important in testing for cutting, core and drilling mud samples and in well-logging interpretation, for location of different oil horizons.

Colour

This is the light transmitted through crude oils. It is yellowish to red for light oils and dark or even opaque for heavy (or low A.P.I gravity) oils.

Some Other Physical Properties

- Odour: This varies greatly in crude oils. High content of light hydrocarbons (paraffins and naphthenes) in crude gives rise to a gasoline-like odour. A pleasant odour is produced if the crude has abundant aromatic hydrocarbons. But with high amount of unsaturated hydrocarbon compounds, sulphur and nitrogen compounds in the oil, it produces a repugnant odour.
- Coefficient of Expansion: This is the measure of volumetric increase of a crude under thermal influence. It increases with increase in A.P.I gravity. Oils containing high amount of (dissolved) gases and possibly of high A.P.I gravities possess high values of coefficient of expansion. Heavy crude oils (low A.P.I gravity) have lower coefficients of expansion.
- Aqueous Solubility: The aqueous solubility of crude oil and its fractions increases linearly with temperature. The rate of solubility becomes significant at temperatures of about 100 °C. At temperatures above 180 °C, crudes occur as molecular solutions in mixed phase with water. At such high temperatures, the nature of the phase enhances primary migration of oil by molecular solution mechanism. And salinity of about 150,000ppm of sodium chloride results in the seperation of liquid hydrocarbons from the aqueous phase.
- Surface tension Effect: Crude oils possess some intermolecular forces of cohesion, expressed as force per unit peripheral outline. Because of this force, oil in dispersed state cannot move through waterwet sand, much less, fine-grained shales. The small forces created by natural hydrodynamic gradients do not overcome those created by surface tension. Consequent the oil is dispersed in the form of globules.
- Flash Point: This is the temperature at which the volatiles rising off the surface of heated oil will ignite with a flash, on passing a flame over the surface. This provides some clew about the gaseous content of the crude oil.

Chemical Properties

The chemical properties of crude oil deal with the chemical nature and the changes in composition in relation to temperature and pressure variations occurring at all times within the oil pool. Some of the chemical properties are related to the origin, migration, and accumulation of the crude oil.

Chemical Nature

Crude oils are composed of liquid hydrocarbon compounds of: paraffin series ranging from pentane to pentadecane (C_5-C_{15}), varying amounts of naphthenes, alkyl benzenes, amomatics, trace amount of semi-solid to solid phase hydrocarbons such as asphaltic matters and waxes. Other secondary constituents are sulphur, nitrogen, oxygen, carbondioxidc, trace metals which include silicon, iron, aluminum, calcium, magnesium, copper, lead, tin, zinc, nickel, molybdenum and vanadium. Various hydrocarbon compounds in crudes have been isolated and analysed at different temperatures ranging from 40 – 180 °C. These hydrocarbons consist essentially of hydrogen and carbon. Some of the compounds are saturated with hydrogen while others are not.

A crude oil is categorized on the basis of the most abundant group of hydrocarbons it contains. A crude oil can be described as paraffin base, naphthene base or mixed base (i.e naphthenic-paraffin base) depending on the percentage composition of its hydrocarbon fractions. The paraffin base crude oils have low density (high A.P.I gravities), and in chemical refineries yield high amount of gasoline and kerosene. The naphthene base crudes produce high amount of lubricating oils, less gasoline and kerosene as the refinery products.

Table Classification of Crude Oils.

Types of Crude	Wax %	Asphalt %	% Composition Of 200 – 300 °C Fraction		
Light Paraffinic	0 – 1.5	0 – 6.0	Paraffin	Naphthenes	Aromatic
			46 – 61	22 – 32	12 – 25
Paraffin Naphthenic	1.0– 6.0	0 – 6	5 – 42	9 – 38	16 – 20
Naphthenic	Trace	0 – 6	15 – 26	61 – 76	8 – 13
Aromatic	0 – 5.0	0 – 20	0 – 8	57 – 78	20 – 35

Hydrogenation of Crude Oils

During the early stages of crude oil formation, there is a remarkable thermal cracking of the organic materials into decomposed complexes, and the coupling activity of anaerobic bacteria processes, and the catalytic influence of such available trace metals as vanadium and nickel, lead, to the transformation of the complex organic matter into alkene rich paraffinic oil. Reservoir catalytic chemical reactions lead to the dissociation of avail sulphides into free sulphur and hydrogen. The elemental hydrogen would convert the alkene rich paraffinic crude oil into an accumulation of gaseous paraffinic oils (of high A.P.I gravity) in relatively close association with the kerogen (or organic source rock).

Paraffin Wax Content

Paraffin waxes in crude oils are semi-solid to solid forms of hydrocarbons, consisting mainly of normal paraffins. These n-paraffins range from about $C_5 – C_{30}$. Admixtures of branched-chain paraffins are also contained. Waxes are complex petroleum substances whose complexity is caused by molecular mixture of branched chain and n-paraffin hydrocarbons, with molecular weights, high enough to be solids at ordinary temperatures.

Paraffin waxes in crude oils have melting points above 30-35 °C. The amount of wax in crude oils varies very greatly. Paraffin base and mixed base crude oils have high amount of paraffin wax up to about 1- 6%. Paraffinic-naphthenic (or mixed) base crude oils usually possess over 10% of paraffin wax, Naphthenic oils contain trace amounts of waxes (usually <1%).

Waxy oils may also contain soft waxes and nonwaxes composed of complex naphthene chains, naphthenic rings and polycyclic aromatic. They are separated by solvent crystallization into viscous oils, soft wax-like materials, and hard, dry microcrystalline solids. High content of paraffin waxes in oils may lead to the clogging of pores of the reservoir rocks. Such oils congeal at atmospheric temperatures and exhibit high pour points. Paraffin waxes are believed to be derived from land plant materials and other terrigenous organic matter of non-marine environments. Paraffin waxes are believed to be derived from land plant materials and other terrigenous organic matter of non-marine environments.

Odd Carbon Chain Lengths

Chemical analysis shows that some crude oils exhibit a detectable predominance of n-paraffins of odd number carbon chains over those of even Members. The odd carbon chains range from C_{17} to C_{33}. The ratio of the sum of the mole percentages of odd carbon n-paraffins to the sum of even Carbon n-paraffins in a specific molecular weight range, serves as an index of odd carbon preference. Crude oils from different environments possess correspondingly different odd carbon preference indices.

Carbon Isotope Ratio

Most crude oils have been proved to contain varying amounts of carbon 13 and carbon 12 isotopes. And the concentration of these isotopes version crudes of different source environments. The ratio of C^{13}/C^{12} isotopes in crude oils is found to be nearly analogous to those of living organic matter. This property thus, establishes a relationship between crude oils and organic sediment. The quantity of isotope carbon-13 in crude oil is significantly different for oils from reservoirs of varying geologic ages.

Sulphur and its Isotope Ratio

In crude oils, sulphur occurs as free sulphur, hydrogen sulphide and in organic sulphur compounds such as thiols and disulphides. Sulphur occurs in trace quantities of about 0.1 to 5.5% by weight in oils. Bacterial reduction processes convert sulphates into hydrogen sulphides and free sulphur. And if the environment is rich in trace metals, available meteoric water would enhance the oxidation of the sulphides into metallic sulphides and hydrogen. These metal sulphides do not influence the composition of the crude. But if the environment is deficient of trace metals there would be direct bacterial oxidation of hydrogen sulphide into elemental sulphur thus producing a sulphur rich oil. High amount of sulphur is found in association with high molecular weight hydrocarbons, (asphaltenes inclusive). High content of sulphur and its compounds corrodes the refinery installations when such crudes are refined.

At temperatures higher than 150 °C, sulphur reacts with some hydrocarbon compounds in the crude to produce thiols and mercaptans. Therefore the presence of elemental sulphur in crude oils suggests a relatively lev: temperature history.

Porphyrins in Crude Oil

Porphyrins are complex hydrocarbon compounds that originate from living organic matters such as chlorophyll and hermins. The basic structure of porphyrins consists of four interconnected rings, each ring containing four carbon atoms and four nitrogen atoms. Porphyrins have high affinity for trace metals of vanadium, nickel, and iron. This property enhances the hybridization of unstable porphyrin substances with nickel or vanadium to produce metalporphyrin complexes.

The latter is produced by the simplification of the ring structures of the chlorophyll matter coupled with the replacement of the magnesium metal atom in the centre, by such trace metals as V, Ni or Fe, thus producing a metallic porphyrin complex. The latter exists more stably in the crude oil, as it offers more resistance to degradation and alteration.

Chemical investigations have shown that paphyrins are one of the important constituents of crude oils. They are commonly associated with isoprenoid hydrocarbon forms; pristane and phytane. These hydrocarbons are derivatives from marine phytoplankton. The long-side chain characteristics of the chlorophyll molecule readily break off to form phytol. And the phytol can either be oxidized to form pristance or be reduced to form phytane, depending on the redox potential of the environment.

Other hydrocarbon groups associated to porphyrins include carotene and terpenoid structures. Porphyrin substances are of moderately low temperature origin, usually destroyed at temperatures of about 200 °C. Thus, their presence in crude oils is an indication that the crude was formed at temperatures below 200 °C.

Trace Metals in Crude Oils

Crude oils contain varying amounts of trace elements some of which include, iron, aluminium calcium, magnesium, copper, lead, tin, astemum, antimony, zinc, silver, nickel, chromium, molybdonium and vanadium. But the most important of these trace elements are vanadium, nickel and iron. The concentration of any of these trace metals is so small that the value is expressed in parts per million. The concentration of trace metals in any crude oil is found to be inversely proportional to the A.P.l gravity of the oil.

Some Other Chemical Properties

Effect of Carbondioxide and Saline Water

High amount of anaerobic microbacteria in some reservoir rocks generate carbondioxide, which in association with hydrocarbon gases, constitute dissolved gases that mobilize the liquid hydrocarbon. The carbondioxide facilitates a decrease in viscosity of the oil and generates internal gas pressure to drive the crude oil from dead-end pockets and through interstitial spaces.

When these gases are dissolved in saline water, sufficient reservoir drive to flow light paraffin base oils is provided. The saline nature of the water reduces its surface tension, thus, creating molecular contact between water and the crudes, subsequently, leading to the effective mobilization of the crude oil.

Geologic and Economic Significance of Properties of Crude Oil

Criteria for Environmental Interpretation

Some properties of crude oils have contributed immensely towards the analysis of the environments of occurrence of the crudes and the understanding of their mode of origin. High wax content in crude oils is a property commonly associated to the genetic environment or the kind of organic matters from which the oil was derived. Paraffin waxes are however conceived to be derived from terrigenous materials (land plants) in the source rock. This is characteristic of paralic to paralic marine environments. Its lithologic sequences are characterized by sandstone and shale intercalations, often waxy oils are associated with inter fingering of continental and marine sediments, and may contain some lenses of coal and oil shales. Researcher thought that waxy oils are of fresh to brackish water origin, occurring in continental, paralic or nearshore environments, examples of waxy crude oils of such environments are those of Nigeria and Eastern Venezuela.

The environment of deposition of source rocks of crude oils can be discerned from the hydrocarbon-sulphur-nitrogen compositional association. Optical activity is a physical property of oils which establishes some relationship between crude oils and marine or near marine environments. This is because of the occurrence of cholesterine substances (source of optical activity) in both exude oils and some marine organisms.

Trace metal complexes occurring in crude oils are considered as important indicators for environmental interpretation. Such trace metals as vanadium, nickel, iron, silicon are believed to have formed from marine organic secretions or from planktonic organisms. And the presence of these elements in crude suggests that the oil must have formed in a marine environment. The presence of iron-porphyrins in crude oils is an indication of source of marine origin. They argued that the characteristic high pH values of about 8.5 for marine environments favour high concentration of iron.

Carbon isotope ratios (C^{13}/C^{12}) in oils are suggestive of the source rock environments. The carbon isotope ratios of crude oils vary with their environments of formation. Researcher found that the average C^{13}/C^{12} ratio for oils of marine environment is greater than 1% whereas the value decreases in oils of nonmarine environment.

Hydrocarbon compounds such as porphyrins and isoprenoid compounds which occur in crude oils contribute immensely to the reconstruction of temperature and paleoenvironraental history of the oils. The ratio of some isoprenoid compounds (pristane-to-phytane), contained in any crude oil is a function of the redox potential of the oil's source rock environment of deposition. Certain relationship has been observed between isoprenoid composition of crude oils and their source rock environment of deposition.

- Peat swamp conditions were characterized by low pH, of 5, low oxygen content and presence of toxic compounds, thus enhancing anaerobic bacterial activity and decarboxylation of phytanic acid. This leads to generation of pristine rich oils characterized by the Pristane-to-heptadecane ratio greater than unity. But the redox potential of the environment does not favour generation of phytane-rich crude oils.
- Source rocks deposited at open water environment generated oils of low pristane/C17 (heptadecane) ratios.
- Source rocks of alternating swampy and open water conditions produce oils of intermediate-Pristane/n-heptadecane ratio.

Criteria for Geologic Interpretations

Optical activity in crudes serves as an important tool for correlation of oils of similar geologic ages. This property depends on the amount of cholesterine substances in the oil. There prevails a transformation of the triterpane and sterane hydrocarbons into polycyclic naphthenes thus, causing a gradual depletion of the concentration of the cholestorine substances with geologic age. Optical rotations of younger crude oils are usually greater than those of older rocks. Researchers measured the optical rotation of several crudes in USSR, and arrived at a conclusion that crude oils of Tertiary age have optical rotations above 1°, while those of Mesozoic and Paleozoic oils were below 1°.

Trace metals are used in geologic interpretation on the basis of the vanadium-to-nickel ratio. Nickel substances in crude oils are more stable than those of vanadium. These metals enter the oil substance very early in its history and leave the substance at unequal rates as the oil matures in its reservoir rock. This gives rise to a net change of the vanadium-to-nickel ratio toward a lower value as the oil becomes older. Researcher noted uniformity in the trend of variations of vanadium nickel, sulphur, resins and increasing asphaltene content in the crude oils of Western Canada. He regarded this trend as a product of alteration. All these characteristics of trace metals in crude oils earn them a recognition as sensitive indicators of alteration and geologic age.

The sulphur isotopic ratio (S^{32}/S^{34}) of crude oils help in the correlation of oils of different source rocks and those of different stages of maturation. Though there occurs some loss of sulphur during maturation and weathering of the crude oils, sulphur isotope ratios show little or no change. Thus, the presence of high sulphur isotopic ratios and generally low sulphur content is indicative of alteration. Very viscous crude oils may suggest some degree of weathering in highly fractured basins.

Criteria for Economic Significance

Some physical and chemical properties of crude oils determine their economic value. Such properties are the A.P.l gravity, wax content, sulphur content and the type of dominant hydrocarbon group (or fractions) of the crude oil. Furthermore, the economic significance of a crude oil also depends on the immediate requirements of the consumer or country. For example light paraffinic (of high A.P.I gravities) oils yield high amount of gasoline and, kerosene as the refinary products. Such countries as India desiring greater supply of kerosene, for domestic source of heat have high demand for crudes that give maximum percentage of kerosene on refining. Crude Oils of moderate gravities (mixed and naphthenic base crudes) are of high demand by advanced countries. This is because of great need for lubricating (industrial) oils and gasolines in such countries for example; U.S.A. makes high demand for both naphthenic and light paraffin oils. Generally however, there is usually a great demand for oils of A.P.I gravities ranging from 20° to 45°. Very light oils (of high A.P.l gravities) and heavy oils (of low A.P.I gravities) can be blended to meet the requirements of the consumers.

The amount of paraffin wax contained in a crude oil influences its economic evaluation. Waxy crude oils are characterized by low calorific values, thus a low capacity to generate heat. This limitation therefore earns any waxy crude oil a relatively low price. Occurrence of sulphur in crude oils is of negative economic significances. This is because; high content of sulphur in crude creates corrosive effects on refining columns. And any installation of special sulphur-resistant steel columns is usually very expensive. The processes of removal of sulphur from crude oils are also expensive and time consuming. Moreover, refining of sulphur bearing crude oils create strong environmental pollution. All these amount to a decrease in demand for, or subsequent reduction in the price of sulphur bearing crude oils. Broadly speaking however, the prices of typical crude oils of all the member countries of OPEC countries are determined by the Organization of Petroleum Exporting Countries (O.P.E.C).

Some Characteristics of Nigerian Crude Oils

A.P.I Gravity

The A.P.I gravities of crude oils of the Niger delta do not exhibit a very consistent trend. The A.P.I gravities of Nigerian crude oils vary from 300 to 400 A.P.I. They change abruptly after a certain depth called the critical depth. The trend of variation of A.P.I gravities of the oils corresponds with the direction of change from medium A.P.I gravity naphthenic oils to light paraffinic oils, as the depth of burial increases. The interface between the heavier and lighter oils is often described as the level of transformation. It is characterized by temperatures of about 60 - 80 °C. The paraffinic oils below the level of transformation were produced by late migration from the source rock.

The accumulation of lighter paraffinic oils at greater depths of burial is as a result of the combination of source and reservoir maturation processes, associated with the high geotemperature conditions at such depths, (which ranges to about 1000 - 115 °C) , and other katagenetic processes.

In some of the oil fields of the Niger delta, extensively faulted reservoirs are thought to have created leakage channels through which inorganic agents such as oxygen and water drain into the reservoir. These inorganic agents cause anomalous weathering of crude oils in the reservoir at high temperatures.

Pour Points

The pour points of crude oils from different oil fields of the Niger Delta show significant variation. This variation reflects a change in the most abundant group of hydrocarbons contained in the crudes. For example, paraffinic waxy oils of the Niger delta possess pour points ranging from 20 to 90 °F. But naphthenic nonwaxy oils of Niger Delta have pour points generally lower than - 25 °F.

Paraffin Wax Content

Most Nigerian oil fields depict a close intercalation of high wax and low wax oils. Some of the paraffin wax bearing crude oils produce as high as 5% wax content. These waxes are believed to be derived from terrigenous (or land plants) materials of the source rock. Researchers supported this view by their successful investigation of the presence of humic, vitrinite and liptinite particles in the Niger Delta source rock. These particles being indicators of terrestrial lithotypes. The widely supported view is that the source beds of the Niger Delta crudes are deeply buried paralic to marine Akata Shale.

Influence of Geotemperatures

The nature of the crude oils in the Niger Delta is being controlled by geotemperatures, thermal maturation and depth of burial. Variation in these elements of katagenesis results in the production of paraffinic-rand paraffin-naphthenic crude oils. API gravities and pour points are low at cooler depths of the reservoir. At relatively higher subsurface temperatures, the crudes are light and characterized by pour points higher than 20 °F. The significance of geoteraperatures on the crude oils of the Niger Delta is not only reflected on the A.P.I gravity, but also in their viscosities and boiling point fractions.

The irregular variation in the properties of crude oils of multiple reservoirs of Niger Delta can also be attributed to random sand bodies that are interbedded with shales, and the influence of tectonic activity, high rates of subsidence and deposition of its source rocks.

Generally speaking, the Nigerian crude oils are light paraffinic and paraffin-naphthenic base crudes. They yield significant amounts of gasoline, kerosene, lubricating oils, and provide naphtha which can be the raw material for fertilizer industries. Some representative properties of Nigerian crudes are shown on table. Nigerian crudes are characterized by little or no occurrence of sulphur. This saves the expense of purchasing sulphur-resistant steel for refining installations, and also saves the time and expensive processes of removal of sulphur in crudes. These favourable factors make the Nigerian crude oils attractive in advanced countries.

Classification of Crude Oil

Natural crude oils are classified according to their density for commercial purposes. The density of crude oil is measured generally in API units. The American Petroleum Institute gravity, or API gravity, is a measure of specific gravity of crude oil compared to water. API values are used to compare densities of various petroleum-based liquids. API gravity is a dimensionless quantity mathematically but is referred to as 'degrees'. The formulae for API is,

$$\text{API Gravity} = \frac{141.5}{\text{specific gravity}} - 131.5$$

The specific gravity of oil fractions is usually determined at 15.6°C. API gravity values of most petroleum liquids range from 10 to 70 degrees. API values of some hydrocarbons are given below. Crude oil is classified as light, medium, heavy, and extra heavy according to its measured API gravity. The API values of crude oil are compared with their specific gravity in Figure below. API gravity greater than 10 indicates that it is lighter than water. API value less than 10 indicates extra heavy crude oil, which sinks under water. As the API gravity of crude oil decreases, it becomes dark.

Table: Types of crude oil as per API value.

	API values	Density (kg/m³)
Light crude oil	API >31.1°	<870
Medium crude oil	22.3-31.1°	870-920
Heavy crude oil	API<22.3°	920-1000
Extra heavy crude oil	API<10°	>1000

On the basis of its sulphur content crude maybe referred to as sweet if it contains relatively little sulphur, or sour if it contains substantial amounts of sulphur.

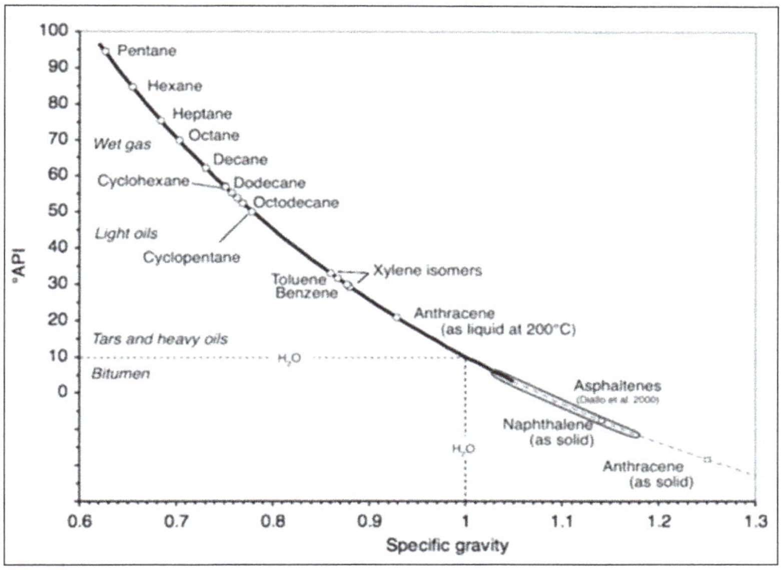

API value of some hydrocarbon compounds and the relationship between API gravity and specific gravity.

Crude oil is refined into useful products like petrol, kerosene, diesel, naphtha, fuel oil and lubricating oil by the method of distillation in a refinery. Crude oil is separated into fractions by fractional distillation. The crude oil fractions at the top of the fractionating column have lower boiling points than fractions at the bottom. All of the fractions are processed further in other refining units.

Benchmark Oils

Benchmark crude oil is the petroleum that serves as a pricing reference for other types of oil and oil-based securities. The benchmark makes it easier for traders, investors, analysts, and others to determine the prices of multiple grades of crude oil varieties and blends. The management of many portfolios will consider prices set by the benchmark crude oil.

Benchmark crude oil primarily tracks these three main types of oil:

1. West Texas Intermediate (WTI) is classified as light sweet crude with a sulfur content of 0.42% or less by weight. The oil is simple to refine into many petroleum products. Production comes from oil fields across the United States, and most refining happens in the Midwest and Gulf South states.

2. North Sea Brent Crude is also described as light sweet oil due to its low sulfur content. This oil comes from reserves in the North Sea. The crude finds usage in the production of gasoline, diesel, heating oil, and several other middle distillate products.

3. Dubai Crude is classified as medium sour crude. The use of the Dubai oil as a benchmark is due to its immediate availability. Dubai oil often provides the benchmark for pricing exports to Asia.

While these three crudes serve as the benchmark for the crude oil industry, many other types of oil from around the world are produced and refined. The selection of a specific crude oil depends on the export and import market, the specific security priced, and other factors.

Petroleum Engineering

A typical workflow for designing, implementing, and executing a project to produce hydrocarbons must fulfill several functions. The workflow must make it possible to identify project opportunities; generate and evaluate alternatives; select and design the desired alternative; implement the alternative; operate the alternative over the life of the project, including abandonment; and then evaluate the success of the project so lessons can be learned and applied to future projects. People with skills from many disciplines are involved in the workflow. For example, petroleum geologists and geophysicists use technology to provide a description of hydrocarbon-bearing reservoir rock. Petroleum engineers acquire and apply knowledge of the behavior of oil, water, and gas in porous rock to extract hydrocarbons.

Some companies form asset management teams composed of people with different backgrounds. The asset management team is assigned primary responsibility for developing and implementing a particular project. Figure below illustrates a hydrocarbon production system as a collection of subsystems. Oil, gas, and water are contained in the pore space of reservoir rock. The accumulation of hydrocarbons in rock is a reservoir. Reservoir fluids include the fluids originally contained in the reservoir as well as fluids that may be introduced as part of the reservoir management program. Wells are needed to extract fluids from the reservoir. Each well must be drilled and completed so that fluids can flow from the reservoir to the surface. Well performance in the reservoir depends on the properties of the reservoir rock, the interaction between the rock and fluids, and fluid properties. Well performance also depends on several other properties such as the properties of the fluid flowing through the well; the well length, cross section, and trajectory; and type of completion. The connection between the well and the reservoir is achieved by completing the well so fluid can flow from reservoir rock into the well. Surface equipment is used to drill, complete, and operate wells. Drilling rigs may be permanently installed or portable. Portable drilling rigs can be moved by vehicles that include trucks, barges, ships, or mobile platforms. Separators are used to separate produced fluids into different phases for transport to storage and processing facilities. Transportation of produced fluids occurs by such means as pipelines, tanker trucks, double-hulled tankers, and liquefied natural gas transport ships. Produced hydrocarbons must be processed into marketable products. Processing typically begins near the well site and continues at refineries. Refined hydrocarbons are used for a variety of purposes, such as natural gas for utilities, gasoline and diesel fuel for transportation, and asphalt for paving.

Petroleum engineers are expected to work in environments ranging from desert climates in the Middle East, stormy offshore environments in the North Sea, and arctic climates in Alaska and Siberia to deep-water environments in the Gulf of Mexico and off the coast of West Africa. They tend to specialize in one of three sub disciplines: drilling engineering, production engineering, and reservoir engineering. Drilling engineers are responsible for drilling and completing wells. Production engineers manage fluid flow between the reservoir and the well. Reservoir engineers seek to optimize hydrocarbon production using an understanding of fluid flow in the reservoir, well placement, well rates, and recovery techniques. The Society of Petroleum Engineers (SPE) is the largest professional society for petroleum engineers. A key function of the society is to disseminate information about the industry.

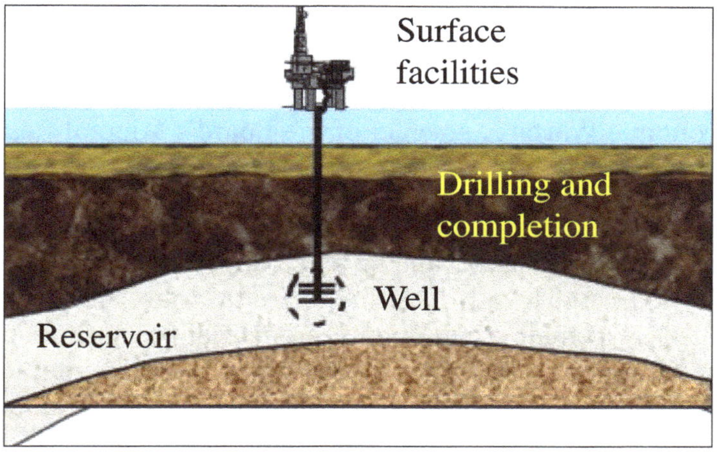

Production system.

Petroleum engineering principles can be applied to subsurface resources other than oil and gas. Examples include geothermal energy, geologic sequestration of gas, and compressed air energy storage (CAES). Geothermal energy can be obtained from temperature gradients between the shallow ground and surface, subsurface hot water, hot rock several kilometers below the Earth's surface, and magma. Geologic sequestration is the capture, separation, and long-term storage of greenhouse gases or other gas pollutants in a subsurface environment such as a reservoir, aquifer, or coal seam. CAES is an example of a large-scale energy storage technology that is designed to transfer off-peak energy from primary power plants to peak demand periods. The Huntorf CAES facility in Germany and the McIntosh CAES facility in Alabama store gas in salt caverns. Off-peak energy is used to pump air underground and compress it in a salt cavern. The compressed air is produced during periods of peak energy demand to drive a turbine and generate additional electrical power.

Theories and Concepts of Petroleum Engineering

Hubbert Peak Theory

A Texas native, Marion King Hubbert obtained his B.S. from the University of Chicago in geology and physics with a mathematics minor in 1926. He would continue his graduate work in the University's Department of Geology, at the time renowned as one of the nation's best, earning his M.S. in 1928, working on a theoretical study of thermodynamic processes contributing to geologic faults. Hubbert garnered applied geophysics experience working as an assistant geologist for the

Amerada Petroleum Corporation. After briefly returning to the University of Chicago to pursue his Ph. D. and work as a teaching assistant, Hubbert went on to accept a faculty position at Columbia University in 1931 where he taught geophysics and completed his Ph. D. in 1937.

Although colloquially best known for his theorization on peak oil, it was during this period that Hubbert published his seminal paper Theory of Ground Water Motion, demonstrating that groundwater flow is determined by both gravity and fluid pressure. The paper provided a physical interpretation of Darcy's Law using a new derivation of the Navier-Stokes equation. Previously thought to remain static in their reservoirs, the publication drastically influenced thinking on how gaseous and liquid hydrocarbons were transported through porous media. During World War II, after leaving his position at Columbia in 1941, Hubbert served as a senior analyst for the United States Board of Economic Warfare. As chief of the Board's Mineral Resources unit, Hubbert oversaw analysis of the Allied war effort's global natural resources, one the most critical of being petroleum.

World War I was vital to establishing, advancing and standardizing the science of estimating petroleum reserves and resources. The advent of motorized warfare shifted the view of petroleum as a commodity supplied by a small group of entrepreneurial prospectors to a national resource fundamental to victory in modern war (for perspective; when Britain entered World War I in 1914 their military possessed only 800 motor vehicles, four years later at the war's end, they had 56,000 trucks and 36,000 cars. In 1916 the United States Geological Survey (USGS) established their Oil & Gas Section. The Section applied methods such as depletion curves to track production and reserves remaining in known fields, and used their data to aid the U.S. military in planning the war effort and also advise the U.S. government on applying taxes to petroleum production. Throughout this time period and through the mid-1940s, the United States accounted for roughly 65% of world petroleum production.

Necessarily, the first attempts at quantifying the nation's reserves, resources and production rates arose from these analyses. In 1919 the USGS Oil & Gas Section estimated domestic petroleum reserves to be 6.74 billion barrels. Chief geologist and head of the Oil & Gas Section David White estimated the supply would last the country 17-18 years extrapolating from current consumption rates. A revised estimate of U.S. petroleum resources, this time including new production from fields in California, was issued in tandem with the American Association of Petroleum Geologists (AAPG) in 1921. Without seismic imaging technology that arose in the mid-1920s, accurate assessment of the nation's oil fields was near impossible, and these early estimates of domestic resources and peak production proved wildly low.

Following his work for the Board of Economic Warfare, Hubbert was employed as a research geophysicist by Shell Petroleum Company from 1943 until 1964, and it was during this time period that he presented his early work on peak oil. In 1949 Hubbert presented his Energy from Fossil Fuels as part of the Symposium on Sources of Energy in Washington D.C., examining annual world production over time for both coal and petroleum. Hubbert's 1949 work focused on the rate of increase of production of both coal and petroleum (neglecting natural gas for want of world production statistics), noting that annual production of coal since 1913 had grown geometrically at a rate of 4% per year since 1913 (implying that annual production doubled every 17 years), while from 1860 to 1929 world crude oil production had grown geometrically at a rate of 9% a year (implying that annual production doubled every 7.5 years).

Pairing these observations with what he deemed "one of the most disturbing ecological influences of recent millennia" that was "the human species proclivity for the capture of energy, resulting in a progressive increase in human population." Hubbert surmised that that the amount of any fossil fuel consumed at any given time would be proportional to the area underneath a curve similar to those which he had presented, concluding:

> "Thus we may announce with certainty that the production curve of any given species of fossil fuel will rise, pass through one or several maxima, and then decline asymptotically to zero. Hence, whole there is infinity of different shapes that such a curve may have, they all have this in common: that the area under each must be equal to or less than the amount initially present."

Hubbert would wait another 7 years to speculate on what this curve might look like, although he would continue the paper by ominously discussing hydropower's ability to meet current energy demand and human existence on an absolute time scale.

In 1956, M. King Hubbert presented Nuclear Energy and the Fossil Fuels before the Spring Meeting of the Southern District Division of Production of the American Petroleum Institute (API). Using petroleum production data from the world, the continental United States and Texas, Hubbert created graphs for each in the same manner as he had done in his 1949 publication, plotting rate of production over time. While initial rates of production initially increased quite rapidly, as had observed for both coal and crude oil in 1949, he noted that this production growth was quite obviously unsustainable in that physical limits prevented production of a finite resource from behaving as such so over any prolonged period. To more effectively extrapolate growth curves for his observed data sets, Hubbert concluded that petroleum depletion could be modeled as function of cumulative production under the premises that:

1. For any production curve of a finite resource of fixed amount, two pointa on the curve are known at the outset, namely that at $t = 0$ and again at $t = \infty$. The production rate will be zero when the reference time is zero, and the rate will again be zero when the resource is exhausted.

2. The second consideration arises from the fundamental theorem of the integral calculus; namely, if there exists a single valued function $y = f(x)$, then,

$$\int_0^{x_1} y dx = A$$

where A is the area between the curve $y = f(x)$ and the x-axis from the origin out to the distance is x_1.

In the case of the production curve plotted against the time on an arithmetical scale, we have as the ordinate,

$$P = dQ/dt$$

where dQ is the quantity of the resource produced in time dt. Likewise, from equation $\int_0^{x_1} y dx = A$ the area under the curve up to any time t is given by,

$$A = \int_0^t P dt = \int_0^t \left(\frac{dQ}{dt}\right) dt = Q$$

where Q is the cumulative production up to the time t. Likewise, the ultimate production will be given by,

$$Q_{max} = \int_0^{\infty} Pdt$$

and will be represented on the graph of production-versus-time as the total area beneath the curve.

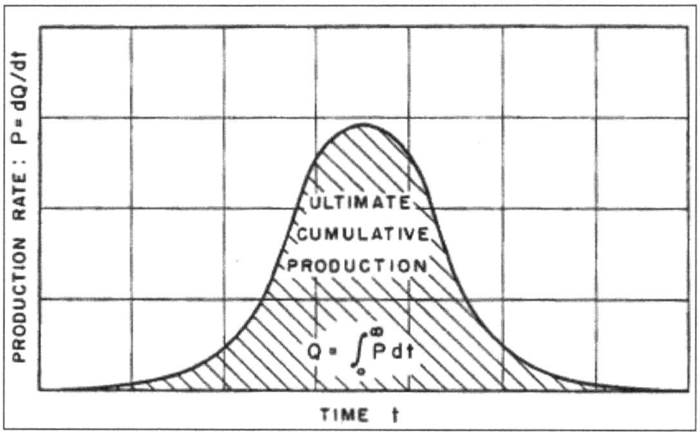

Hubbert's proposed mathematical relationship encompassing the complete production cycle of a finite resource.

The graphical representation of these basic relationships manifested itself in the form of a bell-shaped curve that Hubbert did not define mathematically in his 1956 publication. It is speculated that he drew the curve by hand, calculating the area beneath it to arrive at his original ultimate reserve estimates.

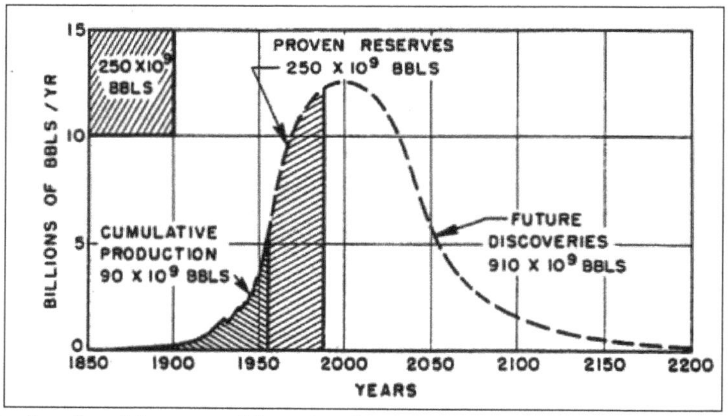

Hubbert's projection for ultimate world crude oil production, assuming a 1250 billion barrel EUR.

Hubbert's curve approach was beautifully simple in that it required only one known variable, Q_{max}, to fit his curve to a data set and extrapolate rates of production over time. The variable Q_{max}, representing ultimate production, is synonymous with Ultimately Recoverable Resources (URR), the approximated cumulative quantity of hydrocarbon that can be economically extracted from a reservoir over its producing lifespan. In his 1956 calculations, Hubbert used a URR of 1250 billion barrels to fit his world production curve. Hubbert fit domestic production curves for both a conservative estimate of 150 billion barrel EUR as well as a slightly more optimistic 200 billion barrel EUR, using an EUR of 60 billion barrels for Texas.

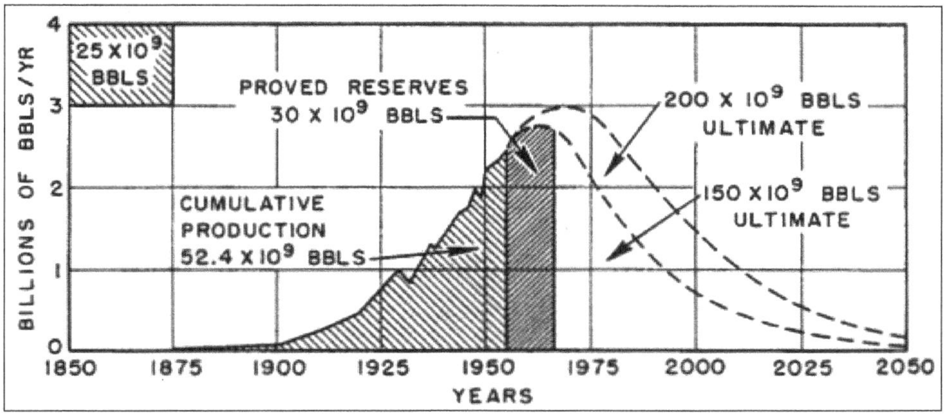

Hubbert's projection for ultimate crude oil production from the Lower 48 United States, the bottom curve assuming a 150 billion barrel URR and the upper curve assuming a 200 billion barrel EUR.

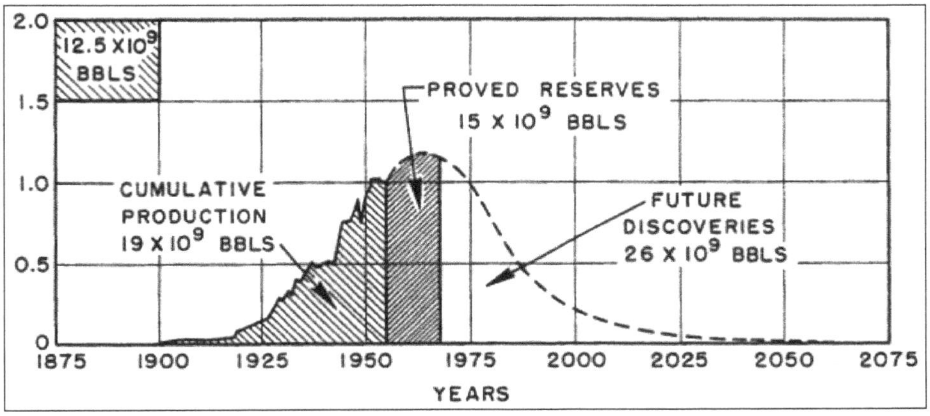

Hubbert's projection for ultimate crude oil production from the state of Texas, assuming a 60 billion barrel EUR.

Hubbert's concept of a moment of peak production implicit to the single curve he had applied to his data sets was not one that had been considered in previous petroleum resource assessments. This would develop into what became known as the theory of peak oil, which by nature of Hubbert's bellshaped symmetric model occurred when exactly half of the ultimate reserve had been produced. His 1956 paper calculated that world peak production of crude oil would occur in the year 2000, with a maximum rate of production of roughly 12.5 billion barrels annually. For U.S. Lower 48 EURs of 150 billion barrels and 200 billion barrels, Hubbert predicted peaks in 1965 and 1970, respectively, with maximum peak production rates of roughly 2.7 billion barrels annually and 3 billion barrels annually.

Hubbert was intrigued by the fact that even when domestic EUR was increased by 33% from 150 billion barrels to 200 billion barrels delayed the occurrence of peak oil by only 5 years. Noting that the versatility and ease of extraction of liquid and gaseous hydrocarbon products had resulted in a rate of production disparate to their magnitude, Hubbert rather sensationally predicted that should the world continue to use fossil fuels as their primary energy source, that "On the basis of the present estimates of the ultimate reserves of petroleum and natural gas, it appears that the culmination of world production of these products should occur within about half a century, while the culmination for petroleum and natural gas in both the United States and the state of Texas should occur within the next few decades."

Hubbert considered peak production to be a critical event in the lifecycle of an exhaustible resource and believed that this point would have significant implications for the manners in which they would be produced and consumed, exacerbated by the increasing demand for energy would come with a growing population. He would touch briefly on the consequences a large energy imbalance held for both domestic industry and national defense purposes before going on to assess alternative energy options as he had done in his 1949 paper, this time focusing on energy from nuclear sources.

Hubbert's Continued Work

Following his initial 1956 work, Hubbert would continue to refine his methods of analysis. Most importantly, in his Techniques of Prediction With Application to the Petroleum Industry presented at the 44th Annual Meeting of the American Association of Petroleum Geologists (AAPG), he assigned mathematical function to the bell curve he had fit to historical data in his earlier work. Hubbert asserted that cumulative production over time could be best approximated by the logistic growth function, with the first derivative of this function, representing annual rate of production, taking the same form as the curve he had drawn in 1956.

Hubbert proposed that magnitude of ultimate reserves could be reached independent of industry or academic expert speculation by examining the relationship between cumulative discoveries, production and proved reserves. At the time of his 1956 publication, Hubbert had declined to make his own ultimate reserve calculations, instead using previously calculated industry estimates to make his preliminary assertions. In the 1950s, best practice in estimating ultimate reserves (used by both the petroleum industry and the USGS) was a method known as volumetric yield analysis. Volumetric yield analysis entailed averaging oil yield produced from a unit volume of sediment in producing basins and assuming that as of yet undrilled, geologically similar basins would produce the same amount per unit volume of sediment to extrapolate ultimate reserves. This form of analysis both assumed comparable yield per unit volume of sediment without scientific justification for doing so, and hinged on subjective measures of geologic similarity. Hubbert realized that his model's dependence on an inherently unreliable variable was a significant flaw, and devised a mathematical means by which to ascertain what he viewed as a more reliable estimate indirectly.

Under the premise that cumulative discoveries (Q_D) for any time t is the sum of oil that has already been produced (Q_P) and oil that has been discovered but not yet produced, (i.e. reserves, represented by Q_R), or:

$$Q_D = Q_P + Q_R$$

Hubbert also ascertained that the curve of cumulative production and the curve of cumulative discovery would be of very nearly identical shape, and have the same asymptote, with cumulative production trailing cumulative discovery with a lag in time that he represented with Δt. With the derivative of this equation with respect to time (t):

$$\frac{dQ_D}{dt} = \frac{dQ_P}{dt} + \frac{dQ_R}{dt}$$

Noting that when proved reserves reach their maximum, their rate of increase would be equal to zero, setting the rate of discovery equal to the rate of production at this time and implying that the ascending rate of production curve would intersect the rate of discovery curve which would have begun to decrease:

$$\text{for } \frac{dQ_R}{dt} = 0, \quad \frac{dQ_D}{dt} = \frac{dQ_P}{dt}$$

With these mathematical foundations in place (graphical interpretations below in Figure), Hubbert had established what would come to be known as the basic tools of modern Hubbert analysis.

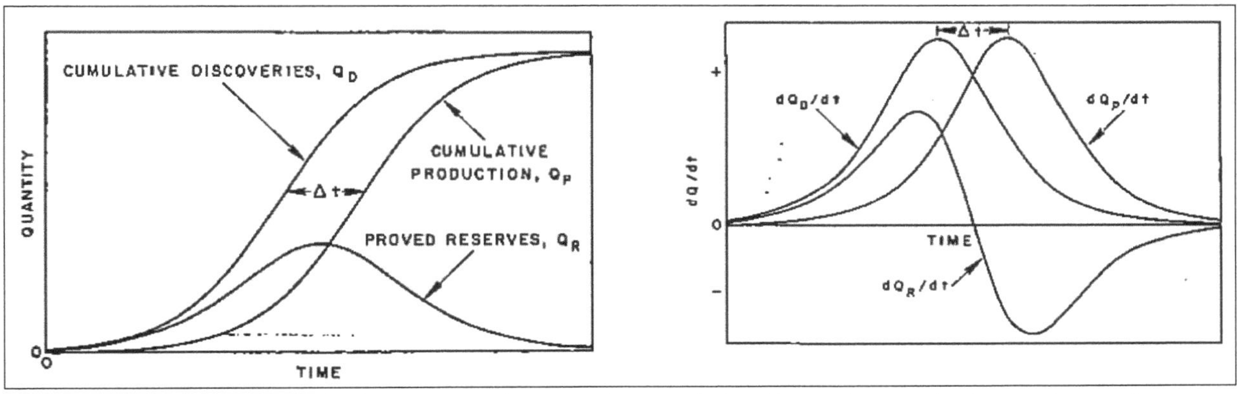

(LEFT) Hubbert's theoretical single-cycle growth curves for cumulative discoveries, production, and proved reserves. (RIGHT) Hubbert's theoretical rate of discovery and production, rate of proved reserve increase.

Hubbert would go on to apply the methods described in his 1959 publication to substantiate his claims in a report submitted to the Committee of Natural Resources in 1962 while serving as chair of the subcommittee on energy. Using crude oil production data from 1860 onwards and estimates of U.S. Lower 48 proved crude oil reserves since from 1937 onward published annually by the API, Hubbert calculated a time lag between U.S. cumulative production and discovery of 10-11 years (best fit for Δt was 10.5), asserting that discovery rate had reached its peak in 1956 while projecting that peak production rate would peak between 1966 and 1967 with ultimate cumulative production of 175 billion barrels.

Considering a contingency allowance of an additional 50 billion barrels, bringing the total to 225 billion barrels, Hubbert claimed this addition would only push peak production to the early 1970s. Hubbert's contentious predictions for peak crude oil production of the U.S. Lower 48 were validated as production rates peaked in 1970 and declined every year subsequently. However, while Hubbert had successfully predicted the timing of U.S. peak oil, his 1962 curve indicating a peak production rate of roughly 2.7 billion barrels annually fell materially short of the actual 1970 production rate of 3.5 billion barrels per year.

Establishing the Parameters of Curve Fitting Techniques and the Uncertainty they Introduce

While it is beyond the scope of this paper to rigorously mathematically justify the use of curve fitting techniques versus other models that have been developed for the analysis of petroleum resource production and consumption, it is important to both acknowledge their limitations as well as establish sound parameters for when their application is both valid and useful.

Ultimately Recoverable Resource (URR)

Ultimately Recoverable Resource (URR) is defined as the sum of cumulative discoveries, future reserve growth at known fields and the volume of oil estimated to be economically recoverable from fields that have not yet been discovered. Inherently, Hubbert's model relies on accurate estimation of URR. Because ultimate production (area under the rate of production curve) delineates projected production over time, it has a larger effect on predictions the model makes than any other variable. URR estimates can be either exogenous or endogenous.

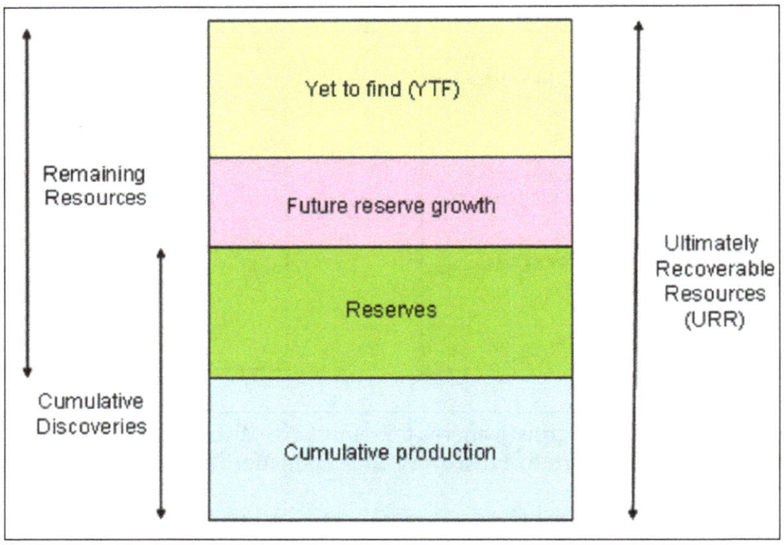

Breakdown of Ultimately Recoverable Resources.

Endogenous Estimates of URR

Hubbert devised his own method of calculating URR so that his models could be independent of unreliable (and in some cases politically motivated) estimates. While at the time this was a reasonable and necessary assumption to make, as methods of geologic estimation have become more certain with improving technology endogenous estimates are increasingly less relied upon. Hubbert's method, which extrapolates peak production rate and time of peak production from the inflection point of the cumulative production curve has been shown to be unreliable.

Exogenous Estimates of URR and Reserve Additions

Exogenous URR estimates are more commonly used in studies employing curve-fitting techniques and act as a limiting external variable independent of the actual curve fitting process. These estimates are generally sourced from national or global geologic authorities such as the USGS or EIA. These estimates present their own aspects of uncertainty to curve-fitting techniques in that historically they have underestimated future additions due to new discoveries and reserve growth. Reserve growth refers to the tendency of estimates for individual fields to increase over time without the addition of new discoveries, and contributes to the majority of reserve additions in most regions of the world, and reserve growth in the Unites States contributed to 89% of proved reserve additions between 1978 and 1990. This effect is attributable to a number of different factors, including improving geologic understanding of an area, improving technology and definitional factors While technological improvements in seismic imaging technology can contribute to reserve growth in older fields where they were not used in original delineation of the play, in the future they will likely mitigate reserve growth attributable to changing geologic understanding of a play as new discoveries will be better understood in the first place.

Improving technology has played a roll in reserve increases due to both other factors as it has improved the accuracy of seismic survey techniques as well as cheapened the extraction process, influencing the definition of what can be referred to as economically extractable. URR estimates are also prone to increase over time as technology improves methods of finding and extracting hydrocarbon, and fluctuate as oil price dictates plays that are economically extractable. While this is potentially problematic, factors that have only long term impacts such as improvements in technology will act more significantly to slow rate of decline after peak production is reached and will have negligible impacts on timing of the peak.

Reserve growth functions have been employed as a means of predicting further reserve growth and rely on the use of measured growth of sample fields to make their forecasts. Typical U.S. 1P data sets exhibit rapid growth immediately after discovery with growth slowing over time, and old fields still record growth after 80 year time periods. Like Hubbert's model, these functions neglect economic variables, but the reason for their growth varies over the production cycle of the play: growth recently after first production is heavily influenced by additions to Original Oil In Place (OOIP), with later additions more commonly attributed to changing recovery factors. Unfortunately, reserve growth functions have been found to be unreliable when projecting potential of a wide range of areas. Calculated rates of reserve growth remain contentious, but it is clear that this effect introduces significant uncertainty to exogenous estimates utilized by curve fitting techniques.

Fractional Distillation

Crude oil is the term for "unprocessed" oil, the materials that comes out of the ground. It is also known as petroleum. Crude oil is a fossil fuel, it was made naturally from decaying plants and animals living in ancient seas millions of years ago - most places you can find crude oil were once sea beds. Crude oils vary in color, from clear to tar-black, and in viscosity, from water to almost solid.

Crude oils are such a useful starting point for so many different substances because they contain hydrocarbons. Hydrocarbons are molecules that contain hydrogen and carbon and come in various lengths and structures, from straight chains to branching chains to rings. Crude oil is processed or refined to produce useable products such as gasoline. The process is very complex and involves both chemical reactions and physical separations. Crude oil is composed of thousands of different molecules.

It would be nearly impossible to isolate every molecule and make finished products from each molecule. Chemists and engineers deal with this problem by isolating mixtures of molecules according to the mixture's boiling point range. The problem with crude oil is that it contains hundreds of different types of hydrocarbons all mixed together. To have anything useful it needs to separate the different types of hydrocarbons. Fortunately there is an easy way to separate things, and this is what oil refining is all about.

Crude Oil Refinery Process

The oil refining process starts with a fractional distillation column. The various components of crude oil have different sizes, weights and boiling temperatures; so, the first step is to separate these components. Because they have different boiling temperatures, they can be separated easily by a process called fractional distillation. After going through the fractional distillation, crude oil is chemically processed to change one fraction into another. Finally, Distillated and chemically processed fractions are treated to remove impurities.

Fractional Distillation

This process is based on the principle that different substances boil at different temperatures. For example, crude oil contains kerosene and naphtha, which are useful fractions (naphtha is made into petrol for cars, and kerosene is made into jet fuel). When you evaporate the mixture of kerosene and naphtha, and then cool it, the kerosene condenses at a higher temperature than the naphtha. As the mixture cools, the kerosene condenses first, and the naphtha condenses later.

Table: Major components of crude oil according to its specific temperature.

Name of the Component	State of matter	Number of carbons	Boiling range	Uses
Residuals	Solid	Multiple-ringed compounds with 70 or more carbon atoms.	Greater than 600 degree Celsius	Coke, asphalt, tar, waxes; starting material for making other products.
Fuel oil	Liquid	Long chain; 20 to 70 carbon atoms.	370 to 600 degree Celsius	Used for industrial fuel; starting material for making other products.
Lubricating oil	Liquid	Long chain; 20 to 50 carbon atoms.	300 to 370 degree Celsius	Used for motor oil, grease, other lubricants.
Diesel distillate	Liquid	Alkanes containing 12 or more carbons.	250 to 350 degree Celsius	Used for diesel fuel and heating oil; starting material for making other products.
Kerosene	Liquid	Mix of alkanes (10 to 18 carbons) and aromatics.	175 to 325 degree Celsius	Fuel for jet engines and tractors; starting material for making other products.
Gasoline	Liquid	Mix of alkanes and cycloalkanes (5 to 12 carbon atoms).	40 to 205 degree Celsius	Motor fuel.
Naphtha	Gas	mix of 5 to 9 carbon atom alkanes.	60 to 100 degree Celsius	Intermediate that will be further processed to make gasoline.
Petroleum gas	Gas	Small alkanes (1 to 4 carbon atoms); commonly known by the names methane, ethane, propane, butane.	40 degree Celsius	Used for heating, cooking, making plastics.

The fractional distillation of crude oil carries out several steps:

- Heating the mixture of the substances of crude oil (liquids) with different boiling points to a high temperature. Heating is usually done with high pressure steam to temperatures of about 1112 degrees Fahrenheit/600 degrees Celsius.

- The mixture boils, forming vapor (gases); most substances go into the vapor phase.
- The vapor enters the bottom of a long column (fractional distillation column) that is filled with trays or plates. The trays have many holes or bubble caps (like a loosened cap on a soda bottle) in them to allow the vapor to pass through. They increase the contact time between the vapor and the liquids in the column and help to collect liquids that form at various heights in the column. There is a temperature difference across the column (hot at the bottom, cool at the top).
- The vapor rises in the column.
- As the vapor rises through the trays in the column, it cools.
- When a substance in the vapor reaches a height where the temperature of the column is equal to that substance's boiling point, it will condense to form a liquid. (The substance with the lowest boiling point will condense at the highest point in the column; substances with higher boiling points will condense lower in the column).
- The trays collect the various liquid fractions.
- The collected liquid fractions may pass to condensers, which cool them further, and then go to storage tanks, or they may go to other areas for further chemical processing.

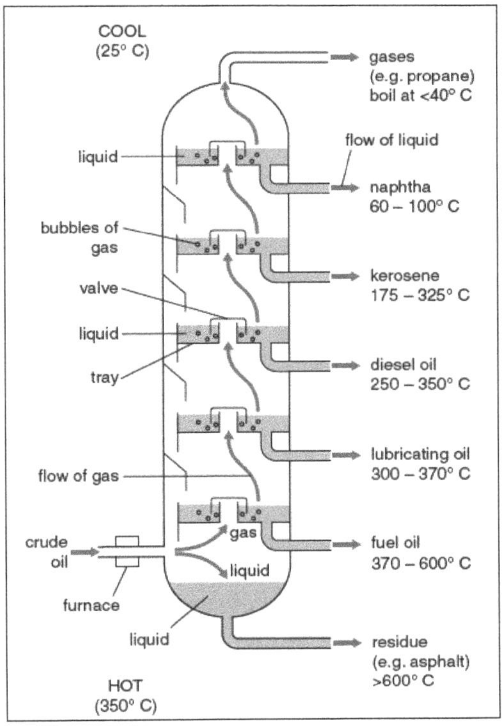

Fractional distillation of Crude oil.

Very few of the components come out of the fractional distillation column ready for market. Many of them must be chemically processed to make other fractions. For example, only 40% of distilled crude oil is gasoline; however, gasoline is one of the major products made by oil companies. Rather than continually distilling large quantities of crude oil, oil companies chemically process some other fractions from the distillation column to make gasoline; this processing increases the yield of gasoline from each barrel of crude oil.

Chemical Processing

Chemical processing of crude oil is to change one fraction into another. Chemical process generally has three methods: (i) Cracking, (ii) Unification and (iii) Alteration.

Cracking

Cracking takes large hydrocarbons and breaks them into smaller ones. There are several types of cracking:

- Thermal: Heats large hydrocarbons at high temperatures (sometimes high pressures as well) until they break apart. Steam - high temperature steam (1500 degrees Fahrenheit/816 degrees Celsius) is used to break ethane, butane and naptha into ethylene and benzene, which are used to manufacture chemicals. visbreaking - residual from the distillation tower is heated (900 degrees Fahrenheit/482 degrees Celsius), cooled with gas oil and rapidly burned (flashed) in a distillation tower. This process reduces the viscosity of heavy weight oils and produces tar, coking - residual from the distillation tower is heated to temperatures above 900 degrees Fahrenheit/482 degrees Celsius until it cracks into heavy oil, gasoline and naphtha. When the process is done, a heavy, almost pure carbon residue is left (coke); the coke is cleaned from the cokers and sold. Photo courtesy Phillips Petroleum Company Catalysts used in catalytic cracking or reforming.
- Catalytic: Uses a catalyst to speed up the cracking reaction. Catalysts include zeolite, aluminum hydro silicate, bauxite and silica-alumina. Fluid catalytic cracking - a hot, fluid catalyst (1000 degrees Fahrenheit/538 degrees Celsius) cracks heavy gas oil into diesel oils and gasoline. Hydrocracking - similar to fluid catalytic cracking, but uses a different catalyst, lower temperatures, higher pressure, and hydrogen gas. It takes heavy oil and cracks it into gasoline and kerosene (jet fuel). After various hydrocarbons are cracked into smaller hydrocarbons, the products go through another fractional distillation column to separate them.

Unification

Sometimes, it needs to combine smaller hydrocarbons to make larger ones; this process is called unification. The major unification process is called catalytic reforming and uses a catalyst (platinum, platinum-rhenium mix) to combine low weight naphtha into aromatics, which are used in making chemicals and in blending gasoline. A significant by-product of this reaction is hydrogen gas, which is then either used for hydro-cracking or sold. A reformer combines chains.

Alteration

Sometimes, the structures of molecules in one fraction are rearranged to produce another. Commonly, this is done using a process called alkylation. In alkylation, low molecular weight compounds, such as propylene and butylene, are mixed in the presence of a catalyst such as hydrofluoric acid or sulfuric acid (a by-product from removing impurities from many oil products). The products of alkylation are high octane hydrocarbons, which are used in gasoline blends to reduce knocking.

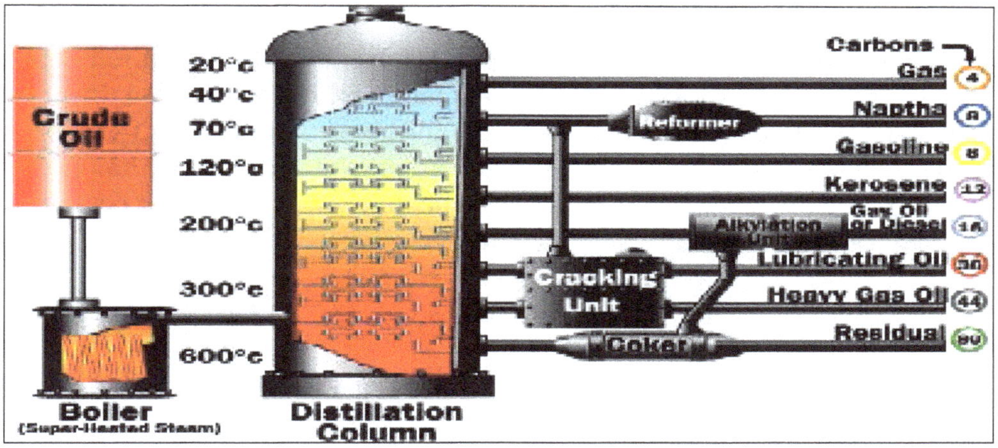

Complete distillation process of Crude oil.

Treating and Blending the Fractions

Distillated and chemically processed fractions are treated to remove impurities, such as organic compounds containing sulfur, nitrogen, oxygen, water, dissolved metals and inorganic salts. Treating is usually done by passing the fractions through the following:

- A column of sulfuric acid removes unsaturated hydrocarbons (those with carbon-carbon double-bonds), nitrogen compounds, oxygen compounds and residual solids (tars, asphalt).
- An absorption column filled with drying agents to remove water.
- Sulfur treatment and hydrogen-sulfide scrubbers to remove sulfur and sulfur compounds.

After the fractions have been treated, they are cooled and then blended together to make various products, such as: gasoline of various grades with or without additives; lubricating oils of various weights and grades; kerosene of various grades; jet fuel; diesel fuel; heating oil; chemicals of various grades for making plastics and other polymers. At the end of it all, Crude oil refinery process is a very complex and lengthy process. Although these refining processes of crude oil exhales large amount of greenhouse gases, we cannot think of our modern civilization without the useful productions from crude oil such as gasoline, jet fuel, fuel for power plants and so many.

Octane Number

Automotive engineers learned that straight-chain paraffins have a much higher tendency to knock than do branched-chain paraffins. The tendency of a particular gasoline to knock is expressed by its octane number (ON).

$$\begin{array}{ll} \text{Heptane} & \text{Iso}-\text{octane} \\ \text{ON}=0 & \text{ON}=100 \end{array}$$

The straight-chain paraffin heptane, C_7H_{16}, is arbitrarily assigned an octane number of 0. The highly branched paraffin formally called trimethylpentane, C_8H_{18}, is assigned an octane number of 100. The name "octane number" derives from the fact that the formal chemical name is abandoned, for convenience, and the compound is referred to as isooctane, and more often simply as 'octane'.

To determine the octane number, the combustion performance of a gasoline is compared in a standardized test engine with various blends of heptane and iso-octane. The octane number of the gasoline is equal to the percentage of iso-octane in the blend that gives the same knock performance as the gasoline being tested. For example, a gasoline with ON=87 has the same engine performance as a blend of 87% iso-octane and 13% heptane. The higher the octane number, the less likely it is that a gasoline will cause engine knock.

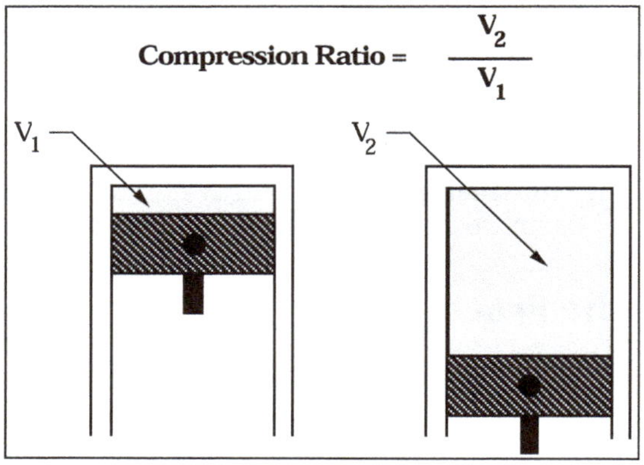

The driving performance of a car depends on the compression ratio of the engine. This is the ratio of the cylinder volume when the piston is at the bottom of its stroke to the volume remaining when the piston is at the top of its stroke. Although, strictly speaking, the ratio would be expressed as a number, say, 8, it is customary to write the compression ratio as 8:1 and refer to it as "eight to one." The efficiency of an engine is proportional to its compression ratio. The higher the compression ratio is, the greater will be the efficiency of conversion of chemical energy of the fuel to kinetic energy of the engine; in other words, the engine's power will be greater. Since power is the rate of doing work, and work must be accomplished to move the vehicle down the road, we experience 'power' as a rapid acceleration of the car and high driving speeds.

If an engine has a high compression ratio, the gasoline/air mixture is compressed to a higher pressure than would be experienced in a different engine of lower compression ratio. Since the gasoline/air mixture is already at a higher pressure in the high-compression ratio engine, it is easier and more likely to be compressed to the pressures that cause knocking during combustion. In other words, the higher the compression ratio of the engine is, the more likely it is to knock with a given grade of gasoline. Consequently, the higher the compression ratio of the engine, the higher the octane number of gasoline that is required to avoid engine knock.

Straight-run gasoline has an octane number of about 55. In contrast, most modern automobile engines require gasolines with octane numbers in the range of 87-93. Not only must refiners increase the total amount of gasoline produced from a barrel of crude, but they must also seek ways to increase the octane number. Three options are available for improving the octane number:

- First, straight-chain paraffins can be converted to branched chain compounds.

- Second, ring-containing aromatic compounds can be blended into the gasoline, because many aromatic compounds have very high octane numbers.
- Third, special antiknock compounds can be added to the gasoline in small quantities.

The use of aromatic compounds in gasoline is on the decline because of Clean Air Act regulations. These compounds tend to contribute much more than paraffins to soot and smoke formation, and with increased concern about air pollution from vehicle exhausts, future gasoline formulations are likely to contain fewer, not more, aromatics. In addition, the simplest aromatic compound, benzene, is thought to be responsible for causing leukemia in laboratory animals, suggesting that long-term exposure to gasoline vapors containing benzene might cause leukemia in humans. The most widely used and most controversial antiknock additive is tetraethyllead (for convenience called 'lead').

The use of tetraethyllead in gasoline is now almost completely discontinued because of environmental concerns. Lead compounds that are produced when tetraethyllead burns in the engine destroy the effectiveness of the catalytic converters in the engine exhaust system. This in turn works against society's efforts to curb the pollution caused by vehicle exhausts. Furthermore, lead compounds leaving the exhaust system of the car get into the environment. Most lead compounds are poisonous to humans and other animals. Thus the best strategy for improving the octane number of gasoline is to convert straightchain paraffins to branched-chain compounds.

In the 1940s, it was discovered that carrying out the cracking operation in the presence of solid materials containing silicon and aluminum oxides, called catalysts, not only broke large paraffin molecules apart into smaller ones, but the carbon chains were also changed from straight to branched structures. A catalyst is a substance that increases the rate of a chemical reaction, but is not itself consumed in the reaction. Many catalysts not only increase the rate of a reaction, but they also allow the overall chemical reaction to be carried out at lower temperatures or pressures by altering the specific molecular processes that take place during the course of the reaction. Essentially, catalysts make it easier to carry out a particular chemical process. The catalytic cracking process in a refinery increases both the total number of gallons of gasoline produced from a barrel of crude oil and the octane number of the product. Catalytic cracking also generally produces less coke than thermal cracking, and wastes less oil. This is today the standard process for improving gasoline yield, and thermal cracking processes are used only to a limited extent.

All crude oils will produce some straight-run gasoline on distillation. This product has too low an octane number for today's cars, but it would be extremely wasteful to throw it away. Straight-run gasoline is, of course, already a gasoline, so it does not need cracking to improve its properties. Rather, all that is needed is for the straight chains of carbon atoms to be rearranged into branched chains or aromatic rings. The shape, and not the size, of the molecules need to be changed. In other words, the molecules need to be 're-formed' and indeed the process by which we do this is called reforming. Like catalytic cracking, reforming is carried out in the presence of a catalyst. The essential difference is that reforming changes only the molecular shape, while catalytic cracking changes both the shape and size of the carbon chains.

Kerosene is used today mainly as a fuel for jet engines and gas turbines. In years past it was widely used for domestic lighting (in kerosene lamps). That market was essentially wiped out by the increasing spread of electric power networks, especially into rural areas, in the first three decades of this century. There is still a small market for kerosene for domestic use, particularly for auxiliary heaters ("space heaters").

Diesel fuel partially overlaps in boiling range with the fuel oils. The boiling range of diesel fuel is 190-380 °C. In contrast to the traditional automobile engines, in which the fuel/air mixture is ignited by a spark plug, diesels have no spark plugs; they rely on extreme compression to ignite the fuel/air mixture. Both automobile and diesel engines are examples of internal combustion engines, meaning that the fuel is ignited and burned inside the engine. We can differentiate the two by referring to the traditional automobile engine as a spark-ignition engine and the diesel as a compression-ignition engine. The compression ratio of a diesel engine is in the range 13:1 to 20:1, whereas a high-performance sparkignition engine might have a compression ratio of 9.5:1; in smaller, economy car engines it might be about 8:1.

Since the diesel engine is a compression-ignition engine, we want (in fact, we need) the fuel to 'knock'. Thus the straight-chain paraffins, which are undesirable in gasoline, are highly desirable components in diesel fuels. The standard of performance of a diesel fuel is the cetane number. The cetane number is determined in analogous fashion to the octane number. In this case, the two reference compounds are hexadecane ('cetane'), $C_{16}H_{34}$, and methylnaphthalene, $C_{11}H_{10}$.

Cetane is assigned a cetane number of 100 and 1-methylnaphthalene, an aromatic compound very unlikely to knock, is assigned a cetane number of 0. Most commercially available diesel fuels have cetane numbers in the range 30-60. For automobiles or light trucks with diesel engines a fuel with cetane number 52-54 is appropriate.

Cetane Methylnaphthalene

Fuel oils have a wide range of uses. Some of the major uses include domestic heating, generation of heat or mechanical power in industrial processes, steam generation in electric power plants, and running the engines on ships. The entire spectrum of fuel oil products sold commercially covers a wide boiling range. Commercial fuel oils are usually assigned a number to indicate the fuel quality.

Table: Classification of fuel oils by numbers.

Number	Name	Color	Calorific value (BTU/gal)
1	Kerosene	Light	137,000
2	Distillate	Amber	141,000
4	Very light residual	Black	146,000
5	Light residual	Black	148,000
6	Residual ("Bunker C")	Black	150,000

Number 2 fuel oil is the common home heating oil. Number 6, or Bunker C oil, is a commonly used industrial heating oil; however, Bunker C usually has such a high viscosity that it must be warmed before it can flow through the fuel system into the furnace. As a rule, the viscosity and the sulfur content of fuel oils increase as the classification number increases. The flash point, the temperature at which the vapors ignite, and the pour point, the lowest temperature at which an oil will still flow, also increase as the number increases.

Global Hydrocarbon Distribution

Oil reserves are an estimate of how much oil can ultimately be recovered, which includes undiscovered or "yet to find" reserves. This broad definition is also called "ultimately recoverable resources." Reserve estimates are based on the probability of finding reserves in certain geological areas. It also assumes new types of technology will make it economically feasible to extract the oil.

Discovered oil reserves is the estimate of future production from known fields. There are three categories. These are based on how likely it is the oil can be recovered using current technology:

- Proved reserves: There is a greater than 90% chance that the oil will be recovered.
- Probable reserves: The chance of actually getting the oil out is greater than 50%.
- Possible reserves: The likelihood of recovering the oil in place is at least 10% but not greater than 50%.

Keep in mind that part of an oil field's probable and possible reserves get converted into proved reserves over time. These discovered reserves are just a small part of the oil in place. It's just not technically feasible to get most of the oil out in any given field.

Proved Reserves

Of the three categories, the most commonly used is proved oil reserves. That's where analysis of geological and engineering data demonstrates with reasonable certainty to be recoverable from known reservoirs. Only the oil that is commercially viable under current economic conditions is counted. If oil prices rise or new technology makes costs lower, then more fields become viable.

Reasonable certainty means that either actual production or conclusive testing has occurred. The testing includes drilling. If not, then the site must be adjacent and similar to areas that have been drilled. The size of the field is determined by the edges where the oil contacts adjacent gas or water formations. Oil is not counted as proved if engineers are uncertain that it can be recovered under current economic conditions. Engineers also don't count it if it's in completely untested areas.

Oil Sands Reserves

The ability to extract oil sands for a reasonable cost has increased the amount of proved reserves. Most of it, totaling about 165 billion barrels, is in Alberta, Canada. The United States imported just over 1.6 billion barrels from these fields in 2019.

Oil sands are sand mixed with a thick substance called bitumen. The bitumen must be heated before it can be used as oil. Two tons of sand must be mined, using three to four barrels of water, to get one barrel of oil. The process is controversial because it uses a lot of energy and water and leaves a scar on the environment that can be seen from space.

How Oil Reserves Work

Reserves are the graveyards of prehistoric plants and tiny marine organisms. Their remains settled at the bottoms of ancient oceans and lakes 541 million to 65 million years ago. Layers of sediment covered them, increasing the pressure and temperature. That changed the chemical composition into oil. Since we use it faster than its being created, oil is considered a non-renewable resource. Most of the big fields in the proved oil reserves are in the Middle East, Venezuela, Canada, and Russia. Here's the number of barrels of proved oil reserves in 2019 for the top 15 countries according to the BP Statistical Review.

Table: 2019 Top 15 Oil Reserves by Country.

Rank	Country	Billions of Barrels	% of World Total
1	Venezuela	303.8	17.5%
2	Saudi Arabia	297.6	17.2%
3	Canada	169.7	9.8%
4	Iran	155.6	9.0%
5	Iraq	145.0	8.4%
6	Russia	107.2	6.2%
7	Kuwait	101.5	5.9%
8	United Arab Emirates (UAE)	97.8	5.6%
9	United States	68.9	4.0%
10	Libya	48.4	2.8%

The list alone doesn't give the whole story, because of the relationships between the countries. Most of them produce more than they use, so they export to importers or those that use more than they produce. They use the U.S. dollars earned from the sale of oil ("petrodollars") to fund government services.

To increase their negotiating power, some oil exporters banded together to form the Organization of the Petroleum Exporting Countries (OPEC) in 1960 to manage world supply and influence prices. OPEC's 13 members hold 80% of the world's proved reserves. The biggest importers of oil in the world are the United States, China, and the European Union. Over the past decade, OPEC has regulated the price of oil to respond to market conditions, including the emergence of shale-based energies.

U.S. Reserves

After years of stagnation, U.S. reserves are now growing again thanks to higher oil prices that make new technologies cost-effective. Horizontal drilling and hydraulic fracturing can extract oil from shale and other "tight" formations or those with very low permeability. Also, the United States maintains the world's largest strategic petroleum reserve at 714 million barrels. The reserve exists to keep the economy running smoothly when there's a crisis or shortage. Since it is not open for production, it's not included as part of the U.S. proved reserves.

Relationship of Reserves to Production

To export a lot of oil, a country must have large reserves; political stability; and expertise to extract, refine, and ship the oil. For example, Venezuela has the world's largest reserves, but political unrest and poor management of the country's oil industry have hampered its production. As a result, it's not even on the list of top 10 oil producers. Since most of the government's revenue depended on oil, the country's economy has descended into chaos. Here are the 2019 top 10 oil-producing countries, according to the BP Statistical Review of World Energy.

Table: 2019 Top 10 Oil Producers by Country.

Rank	Country	Production (mbd)	Share of World Total
1	United States	17.0	17.9%
2	Saudi Arabia	11.8	12.4%
3	Russia	11.5	12.1%
4	Canada	5.7	5.9%
5	Iraq	4.8	5.0%
6	UAE	4.0	4.2%
7	China	3.8	4.0%
8	Iran	3.5	3.7%
9	Kuwait	3.0	3.1%
10	Brazil	2.9	3.0%

Why Oil Reserves Matter

Oil reserves are one of the factors affecting oil prices. Supply, as measured by oil production is also important, and is demand. These factors are reflected in the price of oil futures contracts in the commodities market. They are agreements to buy or sell oil at a specific date in the future for an agreed-upon price. That's why oil prices change daily; it all depends on how trading went that day.

The oil price forecast has shown much volatility because of the changes in oil supply, OPEC's actions, and global demand. Traders look at all three factors, only one of which is reserves. They also look at oil production, which depends on decision-makers in Saudi Arabia, Kuwait, Venezuela, and Russia. The most important factor is demand, particularly from the world's largest user, the United States.

Oil Reserves and Climate Change

When oil is burned, it emits greenhouse gases. Carbon dioxide and other gases act like a blanket that traps the sun's light and heat. Burned oil and other carbon emissions have contributed to a rise in the earth's average temperature of more than 1 degree Celsius since 1880. This global warming has caused the climate to change. Some of its effects are rising sea levels, extreme weather, and ocean acidification.

As a result, there's a growing call to switch global energy supply to renewable energy sources like wind and solar to solve the global-warming crisis. If successful, this movement could make the amount of oil reserves irrelevant.

Explorationists understand a basin and the sequences of tectonic events and various cycles of sedimentation. They unravel the source rock potential and identify reservoirs, traps prospects are identified, and 3D models are prepared using E&P software.

Petroleum System

According to explorationists, a petroleum system is a geologic element and processes that are necessary for the hydrocarbon accumulation. The cross-section below describes the petroleum system along a south Atlantic continental margin. The geologic elements should be present in the order as follows:

- The source rock must contain the organic matter in sufficient amount.
- There must be a porous and permeable reservoir rock for the accumulation of petroleum.
- There must be an impervious sealing cap rock to block the further migration of oil.
- Overburden rocks must buries the source rock to depths where there is optimal temperature and pressure to cock the organic matter for the hydrocarbon generation.

With the extension and faulting of continental crust the rifting of the South Atlantic Ocean has started due to this, the continental crust thinned and eventually split apart. Due to the separation of two parts of the continental crust, the oceanic crust is formed at a mid-oceanic ridge during sea-floor spreading. The continental margin is located where the thinned continental crust meets the oceanic crust. The source and reservoir rocks that were eventually trapped and sealed underneath salt preserve the synrift lacustrine basins. The hydrocarbons that are buried and trapped beneath postsalt marls are migrated from synrift source rocks to limestone reservoirs. The source rocks are also provided by marls. Clayey sandy sediment buried the margin, source rock, reservoir rock, cap rock and overburden during the Tertiary period.

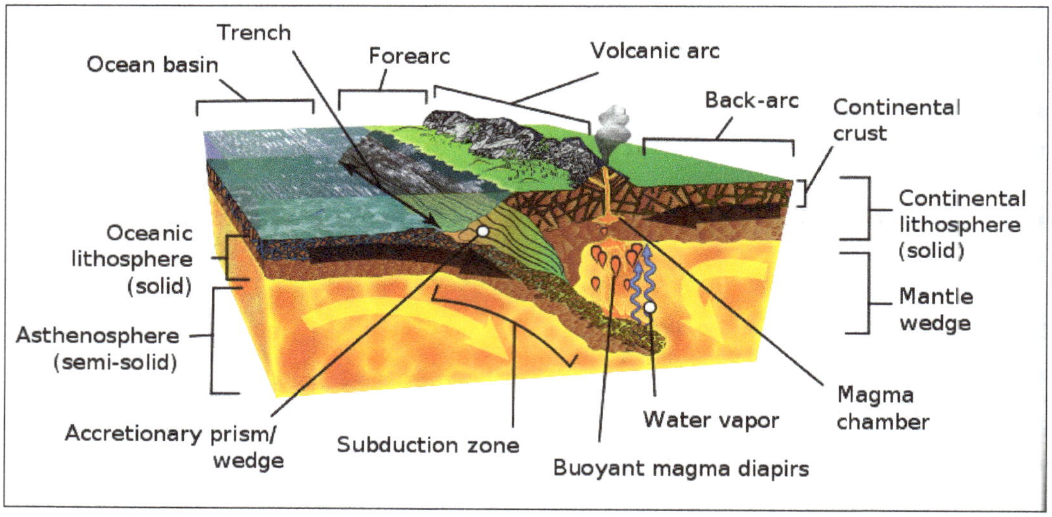

Seismo geologic section along a South Atlantic continental margin rifting of the South Atlantic.

Evaporation in Santos Basin

Between the Rio Grande Rise and the Gulf of Guinea, the salt basins that face one another are among the largest found along Phanerozoic passive ocean margins. The formation of salt basins takes place during the opening stages of the central South Atlantic i.e., during Aptian (125 to 110 Ma). This lower Cretaceous salt deposition appears strikingly similar to the Mid-Late Miocene Red sea (15-5 Ma) in terms of geometric, kinematic and temporal environment. The plates started to separate from one another slowly after the Tristan da Cunha hot spot has started to induce giant volcanic eruptions that covered huge areas of the African-South American lithosphere with thick flood basalts about 143 Ma. Along the newborn plate boundary the narrow rifts, 50 to 80 km (31 to 50m), which has overlapped were found. Basaltic volcanism and anoxic deep-water lakes some deeper than 1,000m (3,300 ft.), similar to lake Tanganyika today- punctuated the geology of such rifts in the Late Hauterivian to Early Barramian (133 to 128 Ma).

There are mainly three conditions necessary for the formation of thick, layered salt deposits are: a basin deep about 1,500 m (4, 900 ft.), a continuous supply of mineral-laden seawater and a warm and arid climate. The basin water level drops quickly and stabilizes to a critical level when the evaporation takes place. The water intake rate is equaled by the evaporation rate. Until the saturation, concentration is reached for the least soluble salt mineral contained in the water the water salinity increases gradually. The halite (rock salt) is followed by layers of calcite, dolomite and gypsum precipitate respectively. The Aptain, about 120 Ma, salt basin was 1,700 km (1,060 mi) long and restricted from open ocean conditions by the Tristan da Cunha hot spot to its south and the embryonic equatorial Atlantic transform margin to its north. The black arrow indicates the direction of the plate movement (Map Courtesy of CR Scotese).

Late Cretaceous Turbidite

The Jubilee oil field offshore Ghana consists of turbiditic reservoir rocks. The Mahogany-I a high-quality pay in an upper Cretaceous turbidite, reservoir confined by the combination of structural-stratigraphic trap is a discovery well encountered at the depth of 90 m (300 ft.). Hyedua-1 well, located 5.3 km southwest of the Mahogany-I discovery, encountered 41 m (130 ft.) in equivalent turbidite sandstone of high- quality reservoir. The water play targeting reservoirs in Late Cretaceous turbidites along the equatorial African transform margin, which stretches from northern Sierra Leone east to southern Gabon in the equatorial segment of the South Atlantic Ocean.

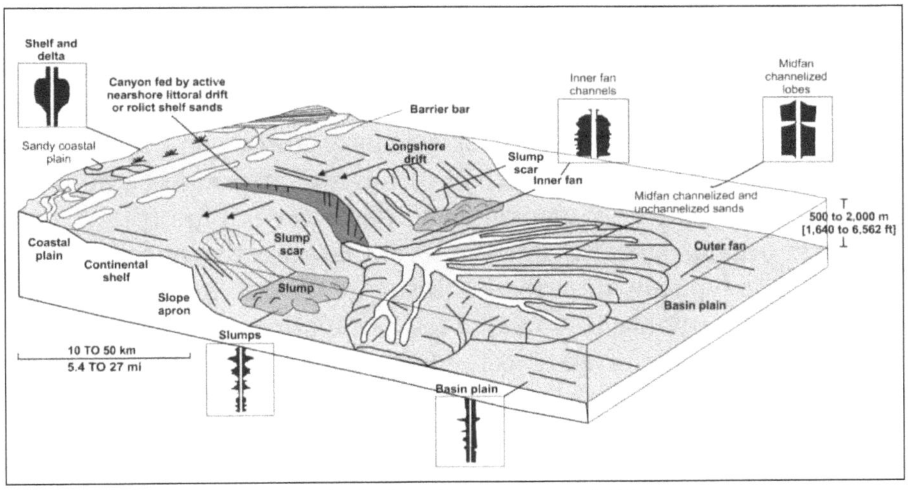

Reservoirs in Late Cretaceous turbidites.

The offshore Ghana fields discovered by deep-water turbidite fields are charged with the hydrocarbons sourced with the rapidly filled organic rich sediments that are active pull-apart basins during the Early Cretaceous epoch. These basins are formed between transform faults formed on rifted continental crust. The beginning of sea floor spreading and the splitting of continents took place during the Albian age. Oblique motion between the two margins is recorded by transform faults and fracture zones, and subsidence and sediment deposition occurred during rifting and subsequent sag of the margins.

In the basin floor that originated from the Guyan continental shelf and slope, explorationists looked for canyons feeding reservoir rocks in channel levee and turbidite fan deposits. Early Cretaceous organic rich sourced shales charged the reservoir rocks that were deposited during continental rifting. These reservoir rocks were sealed and buried by the marine shales. Expected well log responses are plotted for the five types of deposits; the left curve is spontaneous potential or gamma ray, and the right curve is resistivity.

Effects of Oil Spill and Countermeasures

Before discussing the actual effects of an oil spill on various elements of the environment such as birds and fish, the types of effects will be discussed. Toxic effects are classified as acute or chronic, which refers to the rate of effect of the contaminant on an organism. Acute means toxic effects occur within a short period of exposure in relation to the life span of the organism, and are generally the result of exposures to high contaminant concentrations. For example, acute toxicity to fish could be an effect observed within four days of exposure. Chronic means occurring during a relatively long period, usually 10% or more of the life span of the organism, and are generally the result of exposures to low contaminant concentrations. It might take a significant portion of the life span for a chronic toxic effect to be observable, although it could have been induced by exposure to a substance that was normally acutely toxic. Chronic toxicity refers to long-term effects that are usually related to changes in such things as metabolism, growth, reproduction, behavior, or ability to survive.

The effects of exposure to a toxic substance can be lethal or sub-lethal. Lethal effect is typically described in terms of the concentration of the toxicant that causes death to 50% of a test population of the species within a specified period of exposure time (typically four days). This is referred to as the LC_{50}. For example, tests of the effects of various crude oils on Daphnia magna, the water flea, show that 5 to 40 milligrams/liter (mg/L) of the oil for a period of 24 hours is lethally toxic. The units of mg/L are approximately equivalent to parts per million (ppm). Sub-lethal means detrimental to the test organism, but below the level that directly causes death within the test period. Sub-lethal effect is typically described in terms of the con centration of the toxicant that causes an adverse effect or response to 50% of a test population of the species within a specified period of exposure time. This is referred to as the EC_{50}. For example, it has been found that a con centration of 2 ppm of crude oil in water causes disorientation in Daphnia magna when exposed for 48 hours.

Oil can affect animals in many ways, including changing their reproductive and feeding behavior and causing tainting and loss of habitat. Oiling of more highly developed animals such as birds may result in behavioral changes, such as failure to take care of their nests, resulting in the loss of eggs. Even a light oiling can cause some species of birds to stop laying eggs altogether.

Feeding behavior might also change. Seals sometimes react to oiling by not eating, which compounds the negative effects of the oil. The loss of an organism's habitat due to oiling can be as harmful as direct oiling because alternative habitats may not be available, and the animal can perish from exposure or starvation.

Finally, tainting becomes a concern with fish and shellfish after an oil spill. Tainting occurs when the organism takes in enough hydrocarbons to cause an "off" flavor or "off" odor in the seafood. These organisms are unsuitable for human consumption until this taint disappears, which takes from days to up to a year after the exposure, depending on the species. After an oil spill, seafood samples in the area are often tested using both chemical methods and sensory testing. The area is sometimes closed to commercial fishing as a precaution, to prevent direct contamination of fishing gear and catch as well as to maintain public confidence. Loss of confidence in seafood safety and quality can impact seafood markets long after any actual risk to seafood from a spill has subsided, resulting in serious economic consequences.

Oil can enter organisms by several exposure routes: physical exposure, ingestion, absorption, and through the food chain. Animals or birds can come into direct contact with oil on the surface of water, on shorelines, or on land. The effects from fouling of bird feathers or sea otter fur, can lead to mortality from the loss of the ability to thermoregulate in cold waters. Ingestion occurs when an organism directly consumes oil, usually by accident as in the case of birds when oil is ingested as they preen or groom their feathers.

Absorption of volatile components of oil is a common method of expo sure, especially for plants and sessile (immobile) organisms, although it also occurs in birds and mammals. Fresh crude oil has a relative abundance of volatile compounds such as benzene and toluene that are readily absorbed through the skin or plant membrane and are toxic to the organism. After a spill, organisms can also be exposed to oil that passes through several organisms via the food chain. Bioaccumulation rarely occurs because the components of oil are generally metabolized and excreted by fish and mammals or eliminated by shellfish when the exposure to the oil decreases.

The effects of oil on the flora and fauna of a region are influenced by many factors, including the relative abundance (thus how much of the population may be affected), its likelihood of coming into contact with the spilled oil, sensitivity to oil exposure (which varies by season, life stage, physiology, etc.), and how quickly a species population would be able to recover.

Aquatic Environments

When describing at the effects of oil spills, it is convenient to divide aquatic resources into fish, plankton, benthic invertebrates, birds, marine mammals, intertidal fauna, marine plants, and habitats. Although freshwater spills have been less studied compared to marine environments, few differences between similar species have been noted.

Fish

There is often concern about the effect of oil on fish, from both an environmental and a commercial viewpoint, as fish are an important food source. Both pelagic (mid-water) and demersal (bottom-dwelling) fish are exposed primarily through uptake of aromatic hydrocarbons in the water column. The concentration of aromatic hydrocarbons in oils varies as does the toxicity of the different aromatic compounds. Although lethal concentrations are rarely found in open seas, such concentrations can occur in confined waters such as bays and estuaries directly under or near spills. Whereas high concentrations of lighter oil have caused massive fish mortality in some incidents, fish are more typically exposed to sub-lethal concentrations of hydrocarbons. The age of a fish is very important in terms of its sensitivity to hydrocarbons, with adult fish tending to be less sensitive than juveniles. For example, tests have shown that adult salmon are 100 times less sensitive to aromatic hydrocarbons than juvenile salmon. In turn, the juveniles are 70 times less sensitive than the salmon eggs. Several studies have shown that fish larvae or newly hatched fish are often more sensitive than fish eggs. Other variables that determine the toxicity of hydrocarbons are the salinity and temperature of the water, the abundance of food, and the general health of the exposed population.

Oil exposure can cause a range of physiological and pathological changes in fish, some of which are temporary and are not a risk to health or survival. Other sub-lethal effects such as the disruption of growth or decreased assimilation of food may affect long-term survival. Some of the effects noted on fish such as eye cataracts, structural changes of fins, and loss of body weight may be related to the stress of exposure and not directly to the hydrocarbons. In controlled tests, some adult fish species have been found to avoid oil slicks on the surface or dissolved hydrocarbons in the water, but this behavior has not been observed in open water spills. The conclusion is that at least some species would avoid an oil spill on open water, if they are able to escape it.

There is concern that oil spills could disrupt the spawning behavior of anadromous species, such as salmon, that live their adult lives in salt water but return to fresh water streams to spawn. Tests have shown that, while salmon will sometimes avoid oil on open water, the exposure to oil may not badly disrupt their 'homing instinct' as they tend to continue on to their freshwater home streams. Experience in actual spills has not been recorded. There is no evidence that hydrocarbons bioaccumulate in fish or any other aquatic species. Rather, fish and other aquatic organisms tend to 'depurate' or eliminate hydrocarbons that they have taken up.

Fish species that live or spend time close to the water surface, the shore, or the sea floor in shallow water are the most vulnerable to oil spills. Species with eggs or larvae that stay close to the surface and those that feed on organisms near shorelines or on the sea bottom are at greatest risk. Fish that spend most of their life stages in open waters are rarely at risk.

Plankton

Plankton is small plants and animals that live in the water and include phytoplankton and zooplankton. Phytoplankton are microscopic plants such as algae and diatoms that live in the upper layer of the water as they depend on light for photosynthesis. Zooplankton are microscopic animals that feed primarily on phytoplankton. Plankton are important because they are at the bottom of the aquatic food chain. Both phytoplankton and zoo plankton vary in their sensitivity to whole oil or hydrocarbons in the water column. Plankton are killed by relatively low concentrations of oil, but are present in such numbers that lost individuals are replaced quickly with little detectable disturbance. Plankton also tends to eliminate low concentrations of hydrocarbons within days once the exposure ends. Some sub-lethal effects of oil on zooplankton include narcosis, reduced feeding, and disruption of normal responses to light.

Benthic Invertebrates

The benthos refers to the environment on the bottom of bodies of water and includes plankton, fish, and other species. Benthic invertebrates that dwell on or in the sea floor include polychaete worms, bivalves such as clams, and many mobile crustaceans such as crabs, shrimp, lobster, and amphipods. Benthic invertebrates are generally divided into two groups — benthic infauna that reside within the bottom sediments and benthic epifauna that live mostly on the top of the sediments. Mobile forms include starfish, gastropods, and sea urchins. Fast-moving species include amphipods and isopods, tiny invertebrates that are an important food source for fish, bottom-feeding whales, and some species of birds. Mobile species have the advantage of being able to avoid contaminated areas or to quickly re-colonize them, whereas it can take years for sessile (or immobile) organisms to re-colonize an area severely affected by an oil spill.

Benthic species can be killed when large amounts of oil accumulate on the bottom sediments. This can occur as a result of sedimentation, which is the slow downward movement of oil with or without sediment particles attached, or by precipitation down with or in plankton fecal pellets. Sometimes the oil itself is heavy enough to sink. High concentrations of hydro carbons in the water column have killed epifauna, particularly in shallow areas or nearshore environments.

Several trends have been noted in the response of benthic invertebrates to oil. Larval stages are much more sensitive than adults, organisms undergoing molting are very sensitive, and less mobile species are more affected. Sub-lethal hydrocarbon concentrations cause narcosis, (death-like appearance when the organism is not actually dead) in most benthic invertebrates. The invertebrates often recover, although they may be more vulnerable to predators or to being swept away by currents. In 1996, a spill of home heating oil off the US east coast dispersed naturally into a nearshore region. The high level of hydrocarbons caused by dispersion narcotized or killed millions of lobsters which were carried onto the shore where those still alive were killed by stranding. Many other species were also killed including surf clams and other benthic invertebrates. Other sub-lethal effects of oil on benthic invertebrates include developmental problems such as slow growth, differential growth of body parts (deformity), changes in molting times, and occasional anomalies in development of organs. Reproductive effects such as smaller brood sizes and premature release of eggs, reduced feeding, and increased respiration have also been noted in laboratory tests. Benthic infauna will sometimes leave their burrows, exposing themselves to predators. Sea stars have been found to retract their tube feet and lose their hold as a result.

Benthic invertebrates can take up hydrocarbons by feeding on contaminated material and through direct absorption from sediments or water. Most invertebrates depurate (eliminate) hydrocarbons when the water and sediment return to a clean state or if placed in a clean environment. In severe oiling, however, depuration can take months. Sessile (or immobile) species are obviously at a disadvantage and may perish from prolonged exposure to contaminated sediments. Generally, however, all benthic species are affected by a short-term dose of the hydrocarbons in oil.

Birds

Birds are often the most visible biota affected by oil spills, especially in the aquatic environment. The two major pathways of oil exposure for birds are fouling of the feathers and ingestion. Oil contaminates feathers when the birds come into contact with slicks on water or shorelines. For sea birds, this is particularly dangerous because when their feathers are oily, their insulation and buoyancy properties are decreased. Once oiled, a bird rapidly loses its body heat, especially at sea, possibly causing death. Oiled sea birds may stay on land where their temperature loss is not as great. In doing so, however, they are away from their source of food and may die of starvation or predation.

Birds clean their plumage by preening and, in doing so, may ingest some of the oil. Birds may also ingest oil by eating oiled prey. Oil ingestion can result in three categories of toxic effects: (1) reduction in reproduction; (2) destruction of red blood cells leading to anemia; and (3) increased stress resulting in an increased susceptibility to disease, all of which reduce the health and survival of oiled birds.

Contaminated birds may transfer oil to their eggs or young. It has been found that only a few drops of fresh oil can kill the young in an egg. Even when birds ingest only a small amount of oil, they may stop laying eggs or the number of eggs may be reduced. A small amount of oil can also affect the hatchability of the eggs.

Birds that are at most risk of impact from spilled oil include those that: spend much of their time on the water, feed by diving through the surface, form dense flocks during migration and overwintering, concentrate in nesting colonies, or have limited populations (thus listed as endangered or threatened). In many spills, cleaning stations are set up to rehabilitate oiled birds. Although veterinary rehabilitation techniques have improved greatly in the past few years, success rates are still poor as it is very stressful for a wild bird to be captured and handled. Only very sick birds can generally be captured, thus many of the birds brought to the treatment centers are often near death. Survival after cleaning varies by degree of oiling and species, with high survival rates reported for penguins and gulls, moderate survival rates for sea ducks, and low survival rates for shorebirds and pelicans.

Marine Mammals

The effects of oil spills on marine and other aquatic mammals vary with species. Seals, sea lions, walruses, whales, dolphins, and porpoises are: as well as the effects on polar bears and otters. Although these latter two species are not actually marine mammals, they spend much of their time in or near the water. All of these animals are highly visible and cause much public concern when oiled. Seals, sea lions, and walruses are particularly vulnerable to oiling because they haul out, often in very large numbers on the shorelines of small islands, rocks, or remote coasts with few options for new territory. For example, in 2011, 30,000 walruses were hauled out on a 1.6 km of beach in the Chukchi Sea, Alaska. Factors that influence the sensitivity of marine mammals to oil exposure include: whether they rely on fur or blubber for thermoregulation; whether they form dense groups on land or in water; how and on what they feed; and inhalation risks.

External oiling of young seals or sea lions generally causes death because their coats are not developed enough to provide insulation in an oiled state. Oil is often absorbed or ingested and mothers may not feed their young when they are oiled. After a large oil spill in South America, about 10,000 baby seals perished when the beaches of their island were contaminated by oil. Not many adult seals perished at the same site, and those who did probably drowned. Most of the oiled young seals died by starvation after their mothers refused to feed them.

Oiled seals on a rocky shoreline, in this spill in Uruguay, approximately 10,000 seals perished, mostly young seals.

Older seals, sea lions, and walruses can take a large amount of oiling without causing death. If lightly oiled, adult seals survive and the oil is slowly lost. Oiling of both adult and young causes the fur to lose waterproofing and buoyancy. It is not known if seals or their relatives would avoid oil if they could, as this has not been observed at spill sites.

Brief exposure of seals, sea lions, and walruses to volatile oil causes eye irritation and longer exposure can cause more permanent eye damage. Several studies on ingested oil have shown that hydrocarbons accumulate in the blubber, liver, kidney, and other organs, although the levels diminish within a few weeks. Long-term effects have not been observed and are difficult to measure because of the difficulty of approaching relatively healthy seals, sea lions, and walruses.

Whales, dolphins, and porpoises can be exposed to oil in the water column or on the surface when they come up to breathe. Despite this, it has been difficult to determine if these species have been killed as a result of a spill. This is probably due to a number of factors. The most famous case is the disappearance of seven killer whales (out of a stable group of that had been monitored since 1984) shortly after the 1989 Exxon Valdez oil spill in Alaska; by 1990, another six whales disappeared. Thus, the total lost from the group was about 33%. However, during most spills, these species are not considered to be at significant risk. Oil does not adhere to the skins of these mammals and, as they are highly mobile, they are not likely to be exposed for a long period of time. There is little information on the effect of ingested oil on whales and their kin, nor is there any evidence that hydrocarbons would be absorbed from water. However, during the Deepwater Horizon oil spill, dolphins from a coastal bay with heavy and persistent oiling had much higher incidences and severity of disease compared to those from an unoiled area, suggesting that chronic exposures could have long-term effects on coastal dolphins.

Polar bears spend much of their time near or in water, swimming between ice floes hunting seals. The potential for oiling is moderate. It was found that polar bears that are oiled ingest oil through grooming themselves, resulting in death or severe illness. Unfortunately, polar bears are attracted to oil, particularly lubricating oil which they will actually drink. This generally causes temporary illness, but in the case of an oil spill, it could result in death. Few studies have been done of the sub-lethal effects of oil on polar bears as they are difficult to study.

Otters live on or near shorelines and spend much of their time on the water or feeding on shellfish gathered in nearshore waters; they must consume the equivalent of up to one-third of their body weight per day. Oil adheres to the otters' fur causing heat loss, which is the most pronounced effect of oiling. Otters attempt to groom themselves after oiling and thus ingest oil, compounding their difficulty. Oiled otters are often caught and taken to rehabilitation centers for cleaning by trained specialists. If rescued and treated soon enough, some otters can be saved. Such rehabilitation is difficult and expensive, however, as animals may have to be kept for a month before release. In addition, many animals die after their release, possibly as a result of human handling.

Intertidal Fauna

Intertidal fauna include animals that live in the shoreline zone between the high and low tides. These organisms are the most vulnerable to oil spills because they and their habitat are frequently coated during oil spills that strand onshore. Typical fauna include the mobile crabs, snails, and shrimp, sessile (immobile) barnacles and mussels, and sedentary limpets, periwinkles, and tube worms. Heavy oiling will generally kill most species. The area does re-colonize after the spill with the mobile species returning first, but recovery takes months and sometimes years. Re-colonization by plants and sessile species is the major factor in site restoration.

As with other aquatic organisms, light oiling affects the immobile species most and most species will take up oil. Mussels and crabs in particular have been studied for their response to oil. Sub-lethal effects include reduced growth and reproduction rate and accumulation of hydrocarbons. Both mussels and crabs will depurate hydrocarbons when oil exposures decrease. Crabs also show premature or delayed molting. Mussels reduce production of attachment threads, often causing the creature to let go of its hold on its feeding surface. Other intertidal fauna show similar behavior as a result of light oiling.

Shoreline cleaning techniques have a direct effect on the recovery of an intertidal area. Very intrusive techniques such as washing with hot or high-pressure water can remove attached animals and food sources and thus delay recovery, despite removing all of the oil. Intertidal fauna are not highly affected by non-volatile residual oil unless they are coated with it. Recovery is fastest in those areas where oil is removed rapidly after a spill using a non-intrusive technique such as cold water, low-pressure washing.

Marine Plants

Marine plants cover a wide spectrum of plant families and algae. Inter tidal algae, macroalgae, and sea grasses are of particular interest during oil spills. Intertidal algae are an important food source for much of the intertidal fauna and, like the fauna, can be severely affected by an oil spill. Although readily killed by even a moderate oil spill, intertidal algae are usually the first biota to recover after a spill. These algae grow on rock and sediment surfaces and will not re-colonize if these surfaces are still heavily oiled with a light oil. Algae will re-establish on oil-coated rocks if the oil is weathered. Like intertidal fauna, algae are also vulnerable to intrusive cleaning techniques such as washing with hot or high-pressure water. In fact, more algae are killed by these techniques than by oil. Sub-lethal effects include reduced reproduction and respiration rates and changes in color.

Macroalgae include two common groups of plants in North America, Fucus and kelp, both of which include many species and sub-species. Fucus, which often inhabit the lower intertidal and nearshore subtidal zones, are not particularly susceptible to oiling because a mucous coating prevents the oil from adhering to the plant. A heavy oil will cover Fucus, however, and cause death or sub-lethal effects. Kelp generally lives in deeper water and is rarely coated with oil. Both Fucus and kelp will absorb hydrocarbons in the water column, but their effect, including death, depends on the length of time that the concentrations are present.

A dose of a few hours will cause only slight and sub-lethal effects, while a moderate concentration over a few days could cause more serious damage and even death. Both plants will show sub-lethal effects of frond loss, color changes, reproductive slowdown, reduced growth, and accumulation of hydrocarbons. Both plants will also slowly depurate hydrocarbons in clean water. As these plants make up the habitat for complex ecosystems including many forms of animals and other algae, the entire ecosystem can be affected if they are damaged. Recovery for both types of plants and their habitats may take years.

Seagrasses generally inhabit the lower intertidal and nearshore subtidal zones and are extensive in many locations around the world. They are highly productive ecosystems and provide nursery habitat and shelter for many commercially and recreationally important fish and shellfish. Inter tidal seagrasses are sensitive to oil coating of the blades, whereas subtidal seagrasses are exposed via uptake of oil from the water column. Direct coating of the blades can result in bleaching, yellowing, and die-off of the blades, and a reduction in biomass. The beds usually survive as long as the roots are not damaged. However, uptake of hydrocarbons from the water column can cause mortality at high concentrations and sub-lethal effects at low concentrations. Slicks passing over seagrass beds are likely to have short term impacts; greatest impacts have occurred where the beds are exposed to chronic exposures of oil being mobilized from heavily oiled adjacent habitats. Because many of the fauna in seagrass beds are either sedentary (thus cannot escape exposures) or early life stages (which are highly sensitive to oil in the water column), the greatest impacts are to these resources rather than the seagrass bed itself.

Marshes

Manual removal and raking of thick oil and heavily oiled vegetation from a marsh during the Deepwater Horizon oil spill.

Marshes support a rich and diverse flora and fauna, serving as important nesting, breeding, spawning, rearing, and feeding habitats for many species of birds, mammals, reptiles, amphibians, fish, shellfish, and other invertebrates. They also provide many ecological services, including primary production, food-web support, nutrient recycling, water filtration, sediment and storm water retention, shoreline stabilization, storm-surge protection, and soil development. Salt marshes are especially vulnerable to oil spills because they are flooded at high tide and their complex surface traps large quantities of oil. It is also difficult to get into a marsh to assess the damage and clean up the oil.

The effects of oil on a marsh depends on the following factors:

- Oil type is one of the major factors determining the degree and type of impacts on marshes. Lighter oils are more acutely toxic than heavier oils. When spilled offshore, light oils seldom cause extensive damage because they spread into thin slicks. However where large amounts of oil strand on the shoreline, such as large spills in inland waterbodies, small spills in small waterbodies, or spills directly into marshes, the marsh vegetation and fauna can be severely affected.
- Heavy refined oils and most crude oils affect marshes through physical smothering of both leaves and soils. The oil weathering prior to landfall reduces the initial toxicity of the oil.
- The extent of oiling on the vegetation is a key factor. If only the stems or parts of the leaves are oiled, often the marshes recover quickly, usually within one growing season.
- Exposure to waves and currents that speed oil removal is another key factor. The residence time of oil on a shoreline increases as the energy of waves and currents decrease.
- Impacts are more persistent when oil penetrates into the marsh soils or forms a thick layer on the marsh surface. Persistence increases with deeper penetration, with soils high in organic matter, and for sites that are sheltered from natural removal processes.
- Too aggressive cleanup efforts can delay recovery due to trampling from foot traffic, erosion of the marsh surface during flushing, and vegetation cutting under certain conditions.
- Other factors include the degree of contamination of the soils, time of year, and different sensitivities among plant species.

The recovery rate of oiled marshes varies according to these factors. Figure shows the estimated "years to recovery" for 33 spills and field experiments for lightly to heavily oiled marshes. For most spills, recovery occurred within 1–2 growing seasons, even in the absence of any treatment. Recovery is longest for spills in cold climates, sheltered settings, where thick oil is stranded on the marsh surface, spills of light refined oils that are heavily stranded in the marsh, and when intensive cleanup was conducted. Recovery is shortest for spills in warm climates, where there was light to heavy oiling of the vegetation only, where spills were of medium crude oils, and where less intensive cleanup efforts were conducted.

Years to recovery for spills and a few field experiments coded by oil type, from shortest to longest recovery. Shading is used to identify those spills where intensive treatment was conducted. Dashes and question marks are used to represent potential time to recovery based on results of the most recent data.

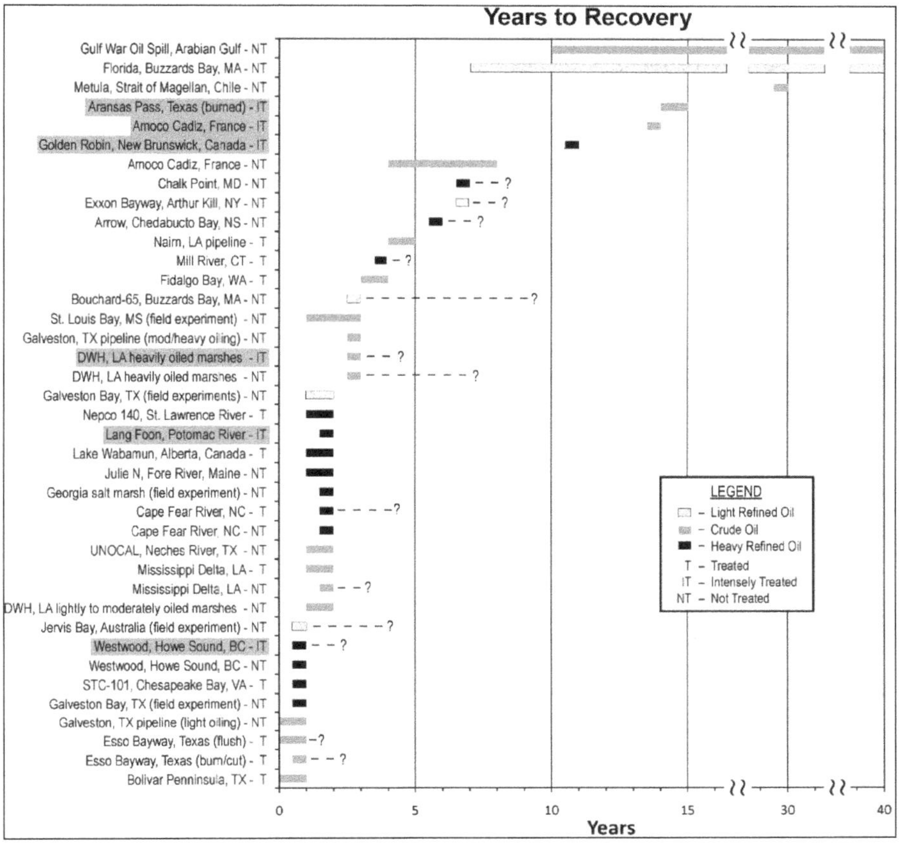

Mangroves

Mangroves are trees and shrubs that grow along much of the shorelines in the tropics. They do not require salt water to grow, but are generally restricted to the intertidal zone because there is less competition with fresh water plants. Like marshes, they provide habitat and a source of food for many important fauna, maintain water quality, and protect the coastal zone from storm damage. As an adaption to growing in water-saturated, low oxygen soils, mangroves have shallow root systems; some species have extensive above-ground roots and small pores that absorb oxygen from the air at low tide.

If these roots are oiled and most of the breathing pores are plugged, the mangrove may die within a few weeks to several months. Impacts from oil spills can last decades because of the longer time for mangroves to reach maturity. Sub-lethal impacts include loss of leaves, production of deformed roots, slower growth, and reduced seedling survival, all of which can reduce the ability of the mangroves to withstand other non-spill stressors. The many other animals supported by a mangrove forest are also at risk from spilled oil, from both direct contact and habitat loss or degradation. Because mangroves are highly susceptible to spilled oil, cleanup options are of very limited effectiveness. Because mangroves have slow recovery rates, they are often priority areas for protection in the event of a spill.

Coral Reefs

Coral reefs occupy a large part of the seas in the tropics of the Pacific and the Caribbean. They are the most diverse and complex marine communities, supporting thousands of fish, algae, and invertebrate species. Studies and actual spills have shown that moderate concentrations of dissolved or dispersed hydrocarbons can kill both the coral and its occupants. Damage depends on the depth, with coral that is near the surface (down to about 6 m) being particularly vulnerable to oil. The time of year is critical because coral reproduction often occurs during a mass spawning event over a few nights, and the early life stages are particularly sensitive to oil. Many coral reef-associated fish and invertebrates have small home ranges and are residents of the reef, thus they may not be able to avoid oil exposures. Many of the animals can repopulate the area relatively rapidly; however, since the coral is their primary support, full recovery depends largely on the recovery or re-colonization of the coral. Once dead, the coral itself can be very slow to recover. Oil also has several sub-lethal effects on coral, such as slowed growth or respiration and unnatural coloration. Infant free-swimming coral are particularly sensitive to oil and have no membrane to prevent oil adsorption into their bodies. Even the smallest amount of dissolved or dispersed oil is lethal to this life form of coral.

Damage Assessment and Restoration

One of the consequences of the Exxon Valdez oil spill was the passage of the Oil Pollution Act of 1990 (OPA 90) in the US. OPA 90 included regulations on how damages to natural resources were to be assessed so as to restore the environment and compensate the public for injury or loss of natural resources and services as a result of the spill and the consequent response actions. These regulations specify a process whereby the injury to natural resources and services is quantified in a way that allows the selection and scaling of various restoration alternatives, which in general include restoration, rehabilitation, replacement, or acquisition of the equivalent resources and/or services.

Restoration can be "primary" — meaning that actions are taken on site to restore the specific injured resource back to what it would be if the spill had not occurred. Examples include planting of vegetation that died as a result of the spill and protection of shorebird nesting sites on oiled shorelines to reduce losses from predators and human disturbances. Restoration can also be "compensatory" — meaning that actions are taken to compensate for the interim losses (usually at another location) while the species or habitat recovers. Examples include restoration of degraded wetlands or creation of new wetlands (preferably within the general area of the oiled wetlands) and protection of summer nesting habitats to increase the population of seabirds that were killed in their wintering area.

Various methods are used to determine both the injury to natural resources and the preferred restoration actions including Habitat Equivalency Analysis (HEA) and Resource Equivalency Analysis (REA). These equivalency methods are based on the principle that compensatory restoration can be achieved through "replacement" of equivalent habitats and resources. These methods avoid having to determine the monetary value of a resource by focusing on cost-effective habitat restoration options. For example, the M/V Athos 1 spilled about 1,000,000 L of a heavy crude oil into the Delaware River near Philadelphia in November 2004. The natural resource trustees (federal and state agencies) conducted studies to determine both the injury to natural resources and appropriate restoration actions to restore these injuries as shown in table. The Damage Assessment and Restoration Plan provides a detailed explanation of how the restoration projects were scaled to compensate for the injured resources). In September 2010, the trustees were awarded US$27,500,000 to support the restoration projects.

Table: Natural Resources Injuries and Restoration Projects for the M/V Athos 1 Spill of 1,000,000 L of Heavy Crude Oil into Delaware Bay.

Natural resources injuries	Restoration projects
Shoreline — 1,729 acres were very lightly, lightly, moderately, or heavily oiled.	Restore 7.0 acres of freshwater tidal wetland to benefit 56 acres within John Heinz National Wildlife Refuge to compensate for tributary losses.
Tributaries — Six tributaries, with a total area of 1,899 acres, were exposed to very light to moderate oiling.	Create roughly 78 acres of oyster reef in the Delaware River to compensate for injuries to aquatic resources, diving birds, and gulls.
Aquatic habitats — 412 acres were exposed to Athos oil.	Remove three dams and a remnant bridge pier; open up an additional 2.6 miles of habitat to anadromous fish, and restore about 10 acres of riparian habitat along the creek edges.
Birds — 11,869 estimated dead (includes direct and indirect losses, a majority of which were swans and geese).	Restore 62.5 acres of degraded wetland and create 35 acres of wet meadow and 100 acres of grassland.
Recreational services — An estimated 41,709 trips on the river were affected by the spill, with an estimated lost value of $1,313,239.	Restore shoreline through the demolition of existing structures, import of fill material, grading of a 0.9 acre site.
	Excavate two shallow wetland ponds in former agricultural areas, convert 16 acres of agricultural lands to cool-season grass pasture, and establish 24 acres of food plots by modifying existing agricultural practices.
	Implement three projects to improve recreational opportunities.

Though natural resource damage assessment is an established process in the US, other nations are also moving towards application of similar approaches. The International Oil Pollution Compensation Funds (IOPC Fund) are three intergovernmental organizations (the 1971 Fund, the 1992 Fund, and the Supplementary Fund) that provide compensation for oil pollution damage resulting from spills of persistent oil from tankers worldwide. The IOPC Fund will pay for what they term "reinstatement of the environment" in addition to the standard costs of property damage, cleanup costs, and economic losses by fishermen and the tourism sector. The IOPC Fund will compensate the costs of measures to reinstate the environment that accelerate the natural recovery process, prevent further damage as a result of the spill, do not result in damage to other resources, are technically feasible, and are not out of proportion to the extent and duration of the damage and the benefits likely to be achieved.

Oil Spill Prevention and Countermeasures

Prevention

Equipment, procedures, and worker training are the three key components of oil spill prevention systems. How they are implemented vary by potential spill sources. Regulated facilities are required by the US Environmental Protection Agency to have Spill Prevention, Control, and Countermeasure (SPCC) plans that prevent oil discharges to navigable waters and adjoining shore lines. Such plans require operators to identify and analyze potential spill hazards and previous spills, identify and ensure availability of resources to remove, to the maximum extent practicable, a worst-case discharge, and describe training, testing, unannounced drills, and response actions of persons on the vessel or at the facility. The US Coast Guard has similar requirements for marine transportation-related facilities.

Double hull tankers are thought to offer a higher level of oil spill prevention because both hulls must be penetrated to cause a release of cargo. A double hull design on average reduced the size of oil spills from tanker vessels by 62% and from tank barges by 20% for spills investigated by the US Coast Guard from 2001 to 2008. Use of single-hull tankers for the transport of oil was phased out worldwide as of 2010 under regulations of the International Maritime Organization, though such tankers may continue to operate until 2015 if they satisfy certain condition assessments. Flag states may grant life extensions to certain types of single hull tankers; however, port states are permitted to deny entry to their ports and offshore terminals to single hull tankers operating under such life extensions. Single hull tankers are prevented from carrying heavy grade oils.

Industry, agencies, and organizations are dealing with three emerging oil spill risks: oil and gas development and maritime transportation in the Arctic; deepwater offshore oil and gas development; and the rapid expansion of rail transportation of crude oil. As the Arctic become increasingly accessible as a result of climate change, offshore oil and gas operations, maritime shipping, and tourism are all expected to increase in the near future. The Arctic is estimated to contain 13% of the world's undiscovered oil and 30% of the world's undiscovered gas reserves. There has been a 118% increase in maritime transit through the Bering Strait between 2008 and 2012. The Emergency Prevention, Preparedness and Response (EPPR) Working Group is one of six working groups of the Arctic Council, which was established in 1996 to foster international co-operation on environmental protection and sustain able development in the Arctic. The EPPR Working Group identified the following recommendations and best practices for the prevention of marine oil pollution in the Arctic:

- Health, Safety and Environment (HSE) Systems: A well-integrated HSE system adapted to Arctic conditions will lead to safer operations. A strong safety culture, characterized by healthy attitudes towards reducing risk and preventing accidents at all levels in the organization is needed.
- Common International Standards: There is a lack of consistency between standards and national requirements, as well as inconsistent enforcement of rules and regulations among Arctic nations. A combination of prescriptive and goal-based requirements is needed.
- Monitoring: An improvement in monitoring methods and technology (i.e. satellites, Unmanned Aerial Vehicles, subsurface vehicles) would provide more accurate data on weather and ice conditions, detect accidents and spill incidents, and deter illegal actions.
- Risk Analysis: Identifying hazards and conducting a risk analysis of the specific operation and implementing risk-mitigating measures are key actions in preventing oil spills.
- Technology Development and Best Practice Sharing: Easy access to information regarding previous experience, best available technologies, best practices, and sharing results of Research and Development projects is of high priority.
- Oil and Gas Facilities: Oil and gas facilities and associated equipment should be adequately winterized, certified for operations in sub-freezing temperatures and potential interaction

with sea ice, and should incorporate leak detection and in-line inspection systems for pipelines.

- Qualified Personnel and Continuous Improvement of Skills: Lack of competence and a need for robust Arctic-specific training requirements were identified as needs to ensure the ability to work in remote areas as well as the continuous improvement of skills.

The risks of deepwater oil and gas exploration and development were demonstrated by the Deepwater Horizon oil spill in 2010, where an estimated 800 million L of oil were released into the Gulf of Mexico during the well blowout. Thus, industry has developed new technologies in blowout prevention systems as well as a focus on safe operating processes, training, and drills. In 2012, BSEE issued rules for Increased Safety Measures for Energy Development on the Outer Continental Shelf (30 CFR Part 250) that:

- Establishes new casing installation requirements.
- Establishes new cementing requirements.
- Requires independent third-party verification of blind-shear ram capability.
- Requires independent third-party verification of subsea blow out prevention stack compatibility.
- Requires new casing and cementing integrity tests.
- Establishes new requirements for subsea secondary blow out prevention intervention.
- Requires function testing for subsea secondary blow out prevention intervention.
- Requires documentation for blow out prevention inspections and maintenance.
- Requires a Registered Professional Engineer to certify casing and cementing requirements.
- Establishes new requirements for specific well control training to include deep water operations.

The use of hydraulic fracturing or "fracking" has dramatically increased US energy production, with crude oil production approaching its highest level in decades. Much of this new oil production is in the mid-continent with the Bakken field in North Dakota that, as of 2014, was producing over 1,000,000 barrels per day. The production of oil from Canadian oils and sis expected to double by 2020. These shifts in crude oil production areas have changed the patterns and means of oil transportation to refineries. Because the capacity of existing pipelines has been strained by this increased production, producers have turned to transport by rail. In 2014, transport of oil by railcar to refineries on the West, East, and Gulf of Mexico coasts is expected to reach 650,000 carloads or roughly 450 million barrels. Efforts to prevent oil spills from rail transportation are focused on tank car design, prevention of derailments, and railroad operations.

New tank car design standards are under development to include an outer steel jacket around the tank car and thermal protection, full-height head shields, and highflow capacity pressure relief valves; there is a plan to aggressively phase out older-model tank cars used to move flammable liquids that cannot be retrofitted to meet these standards, In February 2014, a rail operations safety initiative was agreed to between the railroads and the US government to institute new voluntary operating practices for moving crude oil by rail. These measures include:

- Addition of braking power to allow train crews to apply emergency brakes from both ends of the train in order to stop the train faster.
- Increased track inspections, with at least two high-tech track geometry inspections each year on main line routes over which trains with or more loaded cars of crude oil are moving.
- Use of Rail Traffic Routing Technology to aid in the determination of the safest and most secure rail routes for trains with or more cars of crude oil.
- Reduced train speeds to 40 mph through urban areas.
- Installation of additional wayside wheel bearing detectors.

Countermeasures

Containment in Water

Containment of an oil spill refers to the process of confining the oil, either to prevent it from spreading to a particular area, to divert it to another area where it can be recovered or treated, or to concentrate the oil so it can be recovered or burned. Containment booms are the basic and most frequently used piece of equipment for containing an oil spill on water. Booms are generally the first equipment mobilized at a spill and are often used as long as the oil persists on the water surface. While many pipeline spills occur on land, most of the oil ends up in water bodies and, as such, booms are frequently used.

A boom is a floating mechanical barrier designed to stop or divert the movement of oil on water. Booms resemble a vertical curtain with portions extending above and below the water line. Booms are constructed in sections, usually 15 or 30 meters (m) long, with connectors installed on each end so that sections of the boom can be attached to each other, towed, or anchored. The three basic types of booms are fence and curtain booms, which are common, and shoreline seal booms, which are seldom used.

Booms are used to enclose floating oil and prevent it from spreading, to protect biologically sensitive areas, to divert oil to areas where it can be recovered or treated, and to concentrate oil and maintain an adequate thickness so that skimmers can be used or other cleanup techniques, such as in-situ burning, can be applied. Booms are used primarily to contain oil, although they are also used to deflect oil. When used for containment, booms are often arranged in a U configuration. The U-shape is created by the current pushing against the centre of the boom. The critical requirement is that the current in the apex of the 'U' does not exceed 0.35 m per second (m/s) or 0.7 knots, which is referred to as the critical velocity. This is the speed of the current flowing perpendicular to the boom, above which oil will be lost from the boom by entrainment into the water and under the boom. If used in areas where the currents are likely to exceed 0.35 m/s, such as in rivers and estuaries, booms must be deployed at an angle to the current. The oil can then be deflected out of the strong currents to areas where it can be collected or to less sensitive areas.

Top: Boom failure because of high currents. Bottom: Proper boom deployment that diverts oil out of the stronger currents in the river to an area of slower currents for collection.

If strong currents prevent the best positioning of the boom in relation to the current, several booms can be deployed in a cascading pattern to progressively move oil toward one side of the water-course. This technique is effective in wide rivers or where strong currents may cause a single boom to fail. When booms are used for deflection, the forces of the current on the boom are usually so powerful that stronger booms are required and they must be anchored along their entire length. A boom's performance and its ability to contain oil are affected by water currents, waves, and winds. Either alone or in combination, these forces often leads to boom failure and loss of oil. The most critical factor is the current speed relative to the boom. Failures will occur when this exceeds 0.35 m/s (0.7 knots).

Oil Recovery on Water

Recovery is the next step after containment in an oil spill cleanup operation. Even in the case of land spills, the oil most often flows to a water body from where it is recovered. An important objective of containment is to concentrate oil into thick layers to facilitate recovery. In fact, the containment and recovery phases of an oil spill cleanup operation are often carried out at the same time. As soon as booms are deployed at the site of a spill, equipment and personnel are mobilized to take advantage of the increased oil thickness, favorable weather, and less weathered oil. After oil spreads or becomes highly weathered, recovery becomes less viable and is sometimes impossible.

Skimmers are mechanical devices designed to remove floating oil from a water surface. They vary greatly in size, application, and capacity, as well as in recovery efficiency. Skimmers are classified according to the area where they are used, for example, inshore, offshore, in shallow water, or in rivers, and by the viscosity of the oil they are intended to recover.

A boom and skimmer being used together to collect and recover oil.

Most skimmers function best when the oil slick is relatively thick. The oil must, therefore, be collected in booms or against a shoreline or floating ice before skimmers can be used effectively. The skimmer is placed wherever the oil is most concentrated in order to recover as much oil as possible. Weather conditions at a spill site have a major effect on the efficiency of skimmers. Depending on the skimmer type, most will not work effectively in waves greater than 1 m in height or in currents >0.5 m/s. Most skimmers do not operate effectively in water with ice or debris such as branches, seaweed, and floating waste. Some skimmers have screens around the intake to prevent debris or ice from entering, conveyors or similar devices to remove or deflect debris, and cutters to deal with sea weed. Very viscous oils, tar balls, or oiled debris can clog the intake or entrance of skimmers and make it impossible to pump oil from the skimmer's recovery system.

Skimmers are also classified according to their basic operating principles: oleophilic surface skimmers; weir skimmers; suction skimmers or vacuum devices; elevating skimmers; and submersion skimmers. Over time, all skimming systems become less effective because of the oil's spread into thinner slicks and weathering into more a viscous liquid. Sorbents are materials that recover oil through either absorption or adsorption. They play an important role in oil spill cleanup and are used in the following ways: to clean up the final traces of oil spills on water or land; as a backup to other containment means, such as sorbent booms; as a primary recovery means for very small spills; and as a passive means of cleanup. An example of such passive cleanup is when sorbent booms are anchored off lightly oiled shorelines to absorb any remaining oil released from the shore and prevent further contamination or re-oiling of the shoreline.

Sorbents can be natural or synthetic materials. Natural sorbents are divided into organic materials, such as peat moss or wood products, and inorganic materials, such as vermiculite or clay. Sorbents are available in a loose form, which includes granules, powder, chunks, and cubes, often contained in bags, nets, or socks. Sorbents are also available formed into pads, rolls, blankets, and pillows. The use of synthetic sorbents in oil spill recovery has increased in the last few years. These sorbents are often used to wipe other oil spill recovery equipment, such as skimmers and booms, after a spill cleanup operation. Sheets of sorbent are often used for this purpose.

Sorbent booms are deployed on the water when the oil slick is relatively thin, i.e. for the final "polishing" of an oil spill, to remove small traces of oil or sheen, or as a backup to other booms. Sorbent booms can be placed off a shoreline to recover oil that is mobilized by wave or tidal action; this strategy is often used for marshes where other response options are likely to cause additional harm. They are not absorbent enough to be used as a primary countermeasure technique for any significant amount of oil.

Chemical Agents

Treating the oil with specially formulated chemicals is another option for dealing with oil spills. An assortment of chemical spill-treating agents is available to assist in cleaning up oil. Approval must be obtained from appropriate authorities before these chemical agents can be used. In addition, these agents are not always effective and the treated oil may still have toxic effects on aquatic and other wildlife. Dispersants are chemical spill-treating agents that promote the formation of small droplets of oil that are dispersed throughout the top layer of the water column by wave action and currents. These small oil droplets are less likely to resurface into slicks; instead, they are subject to dilution, transport by currents, and natural weathering process that include dissolution, microbial degradation, and sedimentation. The use of dispersants remains a controversial issue, and special permission is required in most jurisdictions. Generally, in freshwater or land applications, their use is banned.

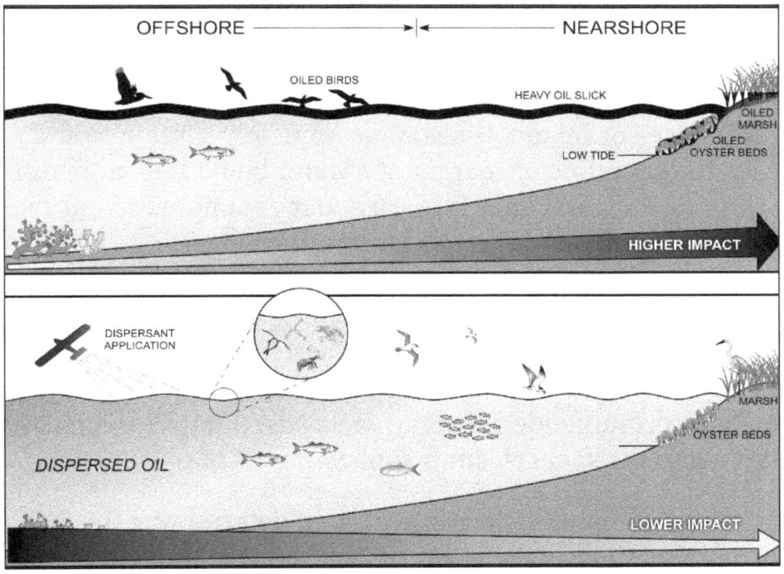

Schematic of the tradeoffs to be considered in the use of chemical dispersants.

Dispersants have a relatively narrow window of effectiveness because oil becomes less dispersible as it weathers, mostly due to an increase in viscosity. There must be sufficient mixing energy from waves or currents, and the wind must not be too strong to prevent the dispersant droplets from landing on the oil slick. There are many restrictions on when and where dispersants may be used as an oil spill countermeasure, including water depth (usually >10 m), distance from shore (often >5 km), season, no spray over birds, sea turtles, or marine mammals, and duration of application. The use of dispersants is considered as part of an trade-off assessment of the impacts of the dispersed oil versus the impacts of the untreated oil slick on sensitive resources. The advantages and disadvantages of the use of chemical dispersants are described in table.

Table: Advantages and Disadvantages of using Chemical Dispersants on an Oil Spill.

Advantages	Disadvantages
Removes surface oil and dilutes oil into the water column	Temporarily increases the bioavailability of toxic oil fractions.
Facilitates natural oil biodegradation.	Temporarily increases localized risks particularly to entrained aquatic species (eggs, plankton, less mobile animals).
Reduces exposure to workers recovering oil at the surface.	Cannot be used under certain weather conditions.
Minimizes impacts on shoreline habitats.	
Minimizes impacts on populations with long life-spans (birds, mammals, turtles).	Cannot be used on all oil types and are less efficient on weathered oils.

In-situ burning is an oil spill cleanup technique that involves controlled burning of the oil at or near the spill site. The major advantage of this technique is its potential for removing large amounts of oil over an extensive area in less or about the same time than other techniques but with the distinct advantage of being a final solution. The technique has been used at actual spill sites for some time, especially on land (including wetlands) and in ice-covered waters where the oil is contained by the ice. During the 2010 Deepwater Horizon oil spill in the Gulf of Mexico, it was used extensively and removed 6% of oil from the water surface.

Shoreline Cleanup

Oil spilled on water is it at sea or inland, is seldom completely contained and recovered and some of it eventually reaches the shoreline or margin of a water body. It is more difficult and time-consuming to clean up shoreline areas than it is to carry out containment and recovery operations at sea. Physically removing oil from some types of shoreline can also result in more ecological and physical damage than if oil removal is left to natural processes. The decision to initiate cleanup and restoration activities on oil-contaminated shorelines should be based on careful evaluation of socioeconomic, aesthetic, and ecological factors. These include the behavior of oil in shoreline regions, the types of shoreline and their sensitivity to oil spills, the assessment process, shoreline protection measures, and recommended cleanup methods. Similarly some of the shoreline types are related to land types and the same cleanup applies.

The type of shoreline or similar land surface is crucial in determining the fate and effects of an oil spill as well as the cleanup methods to be used. In fact, the shoreline's basic structure and the size of material present are the most important factors in terms of oil spill cleanup. There are many types of shorelines, all of which are classified in terms of sensitivity to oil spills and ease of cleanup. The shoreline types in order of increasing sensitivity are: exposed bedrock and man-made solid structures, sand beaches, mixed sand and gravel beaches, gravel beaches, riprap, exposed tidal flats, sheltered rocky shores, peat, sheltered tidal flats, marshes, mangroves, swamps, and low-lying tundra. These types occur on both seashore and freshwater shores.

Many methods are available for removing oil from shorelines or land forms. All of them are costly and take a long time to carry out. The selection of the appropriate cleanup technique is based on the type of substrate, the depth of oil penetration or burial in the sediments, the amount and type of oil and its present form/condition, the ability of the shoreline to support foot and vehicular traffic, the environmental, human, and cultural sensitivity of the shoreline, and the prevailing ocean and weather conditions.

Some recommended shoreline cleanup methods are natural recovery, manual removal, flooding or washing, use of vacuums, mechanical removal, tilling and aeration, sediment reworking, sorbents, and chemical cleaning agents. Surface-washing agents are intended to be applied to shorelines or surfaces to increase the amount of oil released from the surface, or to reduce the pressure and temperature of flushing. Only those agents that cause the released oil to float are used for use on the shoreline, so that the released oil can be recovered. Other agents are essentially industrial cleaners that may be appropriate for cleaning equipment when removed from the water, but not on shorelines.

Numerous guides have been developed to assist responders in selecting the most appropriate cleanup method depending on the type and degree of oiling, the habitat type, and environmental considerations. Table shows a matrix for selection of cleanup methods for marshes.

Table: Recommendations for Response Options in Oiled Marshes by Oil Group.

Oil group descriptions	Oil group				
I — Gasoline products	Response method	I	II	III	IV
II — Diesel-like products and light crudes	Natural recovery	A	A	B	B
III — Medium grade crudes and intermediate products	Barriers/Berms	B	B	B	B
IV — Heavy crudes and residual products	Manual oil Removal/Cleaning DCB B	D	C	B	B
The following codes compare the relative environmental impact of each response method for each oil type:	Mechanical oil removal	D	D	C	C
	Sorbents	—	A	A	B
	Vacuum	—	B	B	B
	Debris removal	—	B	B	B
A = The least adverse habitat impact.	Sediment reworking/Tilling	D	D	D	D
	Vegetation cutting/Removal	D	D	C	C
B = Some adverse habitat impact.	Flooding (deluge)	B	B	B	B
	Low-pressure, Ambient-water flushing	B	B	B	B
C = Significant adverse habitat impact.	Shoreline cleaning agents	—	—	B	B
	Nutrient enrichment	—	B	B	C
D = The most adverse habitat Impact.	Natural microbe seeding	—	I	I	I
I = Insufficient information to evaluate impact or effectiveness.	In-situ burning	—	B	B	B

Sometimes the best response to an oil spill on a shoreline may be to leave the oil and monitor the natural recovery of the affected area. This would be the case if more damage would be caused by cleanup than by leaving the environment to recover on its own. This option is suitable for small spills in sensitive environments and on a beach that will recover quickly on its own such as on exposed shorelines and with non-persistent oils such as diesel fuel on impermeable beaches. This is not an appropriate response if important ecological or human resources are threatened by long-term persistence of the oil.

Cleanup of Oil Spills on Land

When dealing with oil spills on land, cleanup operations should begin as soon as possible. It is important to prevent the oil from spreading by containing it and to prevent further contamination by removing the source of the spill. It is also important to prevent the oil from penetrating the surface and possibly contaminating the groundwater.

Berms or dikes can be built to contain oil spills and prevent oil from spreading horizontally. Caution must be exerted, however, that the oil does not back up behind the berm and permeate the soil. Berms can be built with soil from the area, sand bags, or construction materials. Berms are removed after cleanup to restore the area's natural drainage patterns. Sorbents can also be used to recover some of the oil and to prevent further spreading. The contaminated area can sometimes be flooded with water to slow penetration and possibly float oil to the surface, although care must be taken not to increase spreading and to ensure that water-soluble components of the oil are not carried down into the soil with the water. Shallow trenches can be dug as a method of containment, which is particularly effective if the water table is high and oil will not permeate the soil. Oil can either be recovered directly from the trenches or burned in the trenches. After the cleanup, trenches are filled in to restore natural water levels and drainage patterns.

Suction hoses, pumps, vacuum trucks, and certain skimmers and sorbents are generally effective in removing excess oil from the surface, especially from ditches or low areas. The use of sorbents can complicate cleanup operations, however, as contaminated sorbents must be disposed of appropriately. Sorbents are best used to remove the final traces of oil from a water surface. Any removal of surface soil or vegetation also entails replanting.

Mechanical recovery equipment, such as bulldozers, scrapers, and frontend loaders, can cause severe and long-lasting damage to sensitive environments. These devices can be used in a limited capacity to clean oil from urban areas, roadsides, and possibly on agricultural land. The unselective removal of a large amount of soil leads to the problem of disposing of the contaminated material. Contaminated soil must be treated, washed, or contained before it can safely be disposed of in a landfill site. This could cost thousands of dollars per ton.

References

- Introduction-to-Petroleum-Engineering: um-palembang.ac.id, Retrieved 17, May 2020
- Peebles-Senior-Essay: earth.yale.edu, Retrieved 23, January 2020
- Distillation-process-of-Crude-oil-261551891: researchgate.net, Retrieved 25, July 2020
- Oil-reserves-definition-categories-world-s-largest-3305873: thebalance.com, Retrieved 10, April 2020
- Oil-Spills-Causes-Consequences-Prevention-and-Countermeasures-301932090: researchgate.net, Retrieved 28, June 2020

2

Petroleum Exploration and Extraction

The search for deposits of petroleum and natural gas in the Earth using petroleum geology is known as petroleum exploration. Petroleum extraction is the process by which petroleum is extracted from the deposits beneath Earth's surface. This chapter discusses in detail the various methods of petroleum extraction like oil well, drilling technique, mud logger, drilling rig and deep well drilling, etc.

Petroleum Exploration

The role of exploration is to provide the information required to exploit the best opportunities presented in the choice of areas, and to manage research operations on the acquired blocks. An oil company may work for several years on a prospective area before an exploration well is spudded and during this period the geological history of the area is studied and the likelihood of hydrocarbons being present quantified.

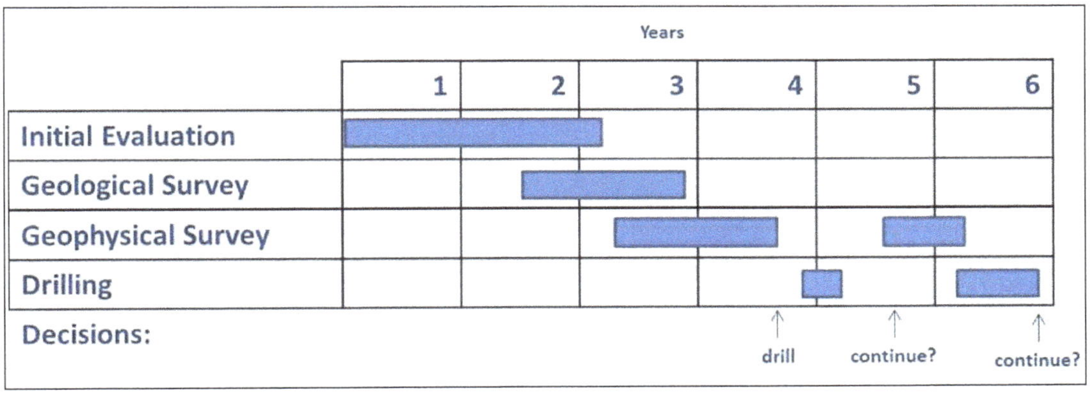

Stages of a typical exploration program.

Exploration is responsible for handling the risk intrinsic in this activity, and this is generally achieved by selection of a range of options in probabilistic and economic terms. Indeed, exploration is a risk activity and the management of exploration assets and associated operations is a major task for oil companies. The risk cannot be eliminated entirely but can be controlled and reduced adopting appropriate workflow, conceptual and technological innovations. When it's been decided to start up with an exploration project in a basin or in a larger area containing several basins, the quantity and quality of available data must be acquired and evaluated – geological data, type of reserves, production of existing fields (if any), etc. Basin assessment/evaluation is the first step to undertake the study of the area under interest. Technological development has provided oil companies with Basin Modeling – which is a numerical simulations that allows the temporal reconstruction of the history of a sedimentary basin and the associated evolution of the processes related to the formation of petroleum accumulations.

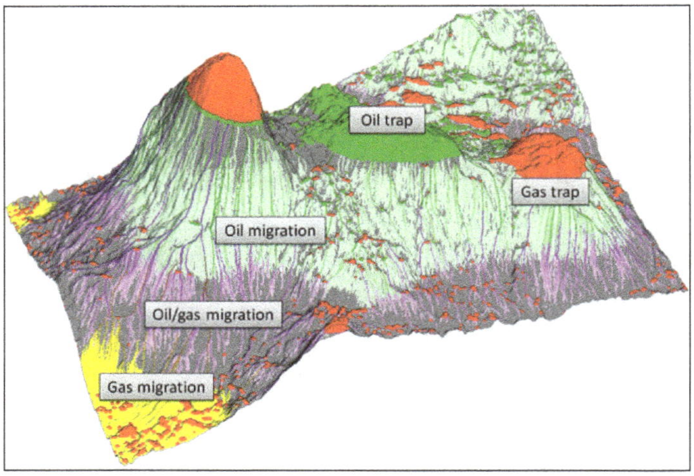

Basin modeling – Petroleum system.

On the basis of data and evidences collected from the preliminary studies, the company management, in the light of the possibilities and the probabilities of a discovery based on G&G data, aside from considerations of an economic nature, may decide to move to the following stage, which is the acquisition (through direct negotiations or by taking part in bids, etc.) of the legal right to perform prospecting in the selected area/block.

The owner of the mining right is normally the State, with which the oil company stipulates a contract establishing the contracting parties' rights. Production Sharing Contracts and service con-tracts are frequently adopted nowadays. The sequence of activities covered by an exploration permit is fairly uniform, and includes:

- The creation of a database.
- The analysis of available data.
- The programming of mapping and geological and photo-geological surveys.
- Seismic surveys and interpretation of seismic data.
- The choice of well locations, drilling.
- The analysis of results and the decision as to whether or not to proceed with the application for a lease or to release the area after fulfilling obligations.

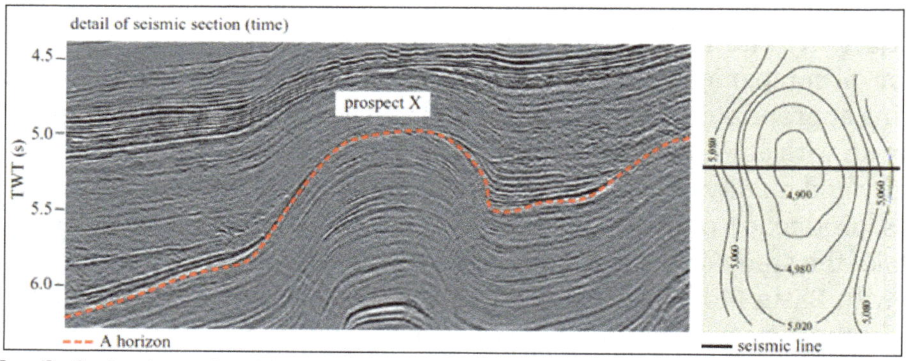

A detail of seismic section (time) and A horizon depth map (meters) referred to prospect X.

Goal of exploration is to identify and locate a prospect, to quantify the volume of hydrocarbon which might be contained in the potential reservoirs and to evaluate the risk inherent the project itself. A prospect is a viable target evidenced by geological and geophysical indications that is recommended for drilling exploration well.

The prospects identified must be technically practicable and meet the market conditions to guarantee a financial return on investments. The results obtained by drilling the exploratory wells indicate whether the initial geological hypotheses are correct or whether variations are found.

All this will allow the fine-tuning of the economic analysis of the project possibly turning hypothetical reserves into proved ones. Where profitability does not meet the standards of the company, it leads to the termination of further investments.

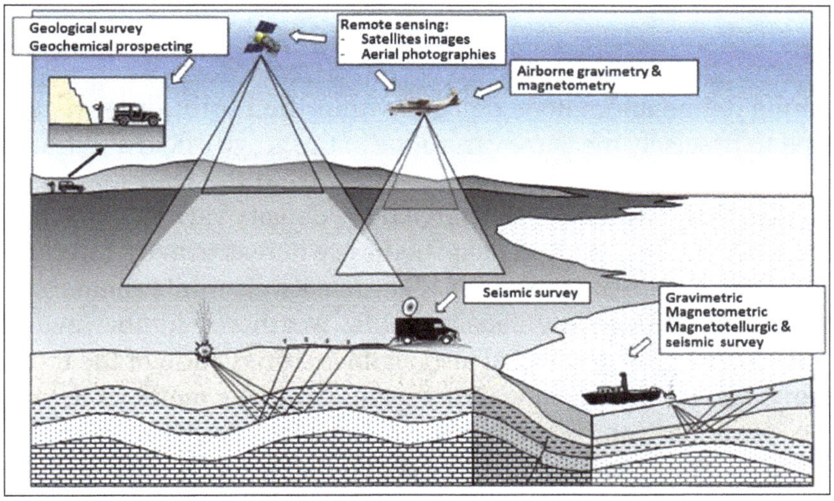

The main petroleum exploration techniques.

Geological Mapping and Prospecting

Geological mapping and prospecting are valuable techniques in an petroleum exploration.

- Geological mapping: It is basically a technique which allows a graphical presentation of geological observations and interpretations.
- Geological prospecting makes use of geological disciplines such as petrography, stratigraphy, sedimentology, structural geology, and geochemistry.

Such disciplines are used to achieve different targets but it must be stressed that their integration is fundamental to depict a picture of reality.

Geophysical Methods

Depth of burial (of contaminant plumes, for instance) is extremely important in assessing the potential applicability of geophysics. Geophysical responses for more deeply buried sources decrease in amplitude and increase in spatial wavelength until they disappear into geologic noise. Physical properties of cover, host rock, and mineral waste strongly influence responses of potentially hazardous material and are also important for evaluating the utility of a method in geo-environmental investigations. Some geophysical methods, such as gamma-ray spectrometry and remote sensing, measure surface attributes; others, such as thermal and some electrical methods are limited to detecting relatively shallow features but may help identify features at greater depth. Secondary effects of deeper features, such as geochemical haloes, can often be identified by these methods. Geophysical modeling provides generalized and non-unique solutions to questions concerning the geometry of subsurface geologic relations. The non-uniqueness of these solutions is both a mathematical problem and one related to the multiplicity of sources that can cause geophysical anomalies. This feature is an implicit uncertainty in the discussion that follows. Environmental geophysics, like exploration geophysics, requires complimentary geophysical surveys integrated with geochemical and geologic insight.

Types of Geophysical Methods

Gravity Method

Gravity measurements define anomalous density within the Earth; in most cases, ground-based gravimeters are used to precisely measure variations in the gravity field at different points. Gravity anomalies are computed by subtracting a regional field from the measured field, which result in gravitational anomalies that correlate with source body density variations. Positive gravity anomalies are associated with shallow high density bodies, whereas gravity lows are associated with shallow low density bodies. Thus, deposits of high-density chromite, hematite, and barite yield gravity highs, whereas deposits of low-density halite, weathered kimberlite, and diatomaceous earth yield gravity lows. The gravity method also enables a prediction of the total anomalous mass (ore tonnage) responsible for an anomaly. Gravity and magnetic methods detect only lateral contrasts in density or magnetization, respectively. In contrast, electrical and seismic methods can detect vertical, as well as lateral, contrasts of resistivity and velocity or reflectivity.

Applications of gravity to mineral deposit environmental considerations include identification of lithologies, structures, and, at times, ore bodies themselves. Small anomalous bodies, such as underground workings, are not easily detected by gravity surveys unless they are at shallow depth.

Magnetic Method

The magnetic method exploits small variations in magnetic mineralogy (magnetic iron and iron-titanium oxide minerals, including magnetite, titanomagnetite, titanomaghemite, and titanohematite, and some iron sulfide minerals, including pyrrhotite and greigite) among rocks. Measurements are made using fluxgate, proton-precession, Overhauser, and optical absorption magnetometers. In most cases, total-magnetic field data are acquired; vector measurements are made in some instances. Magnetic rocks contain various combinations of induced and remanent magnetization that perturb the Earth's primary field. The magnitudes of both induced and remanent magnetization depend on the quantity, composition, and size of magnetic-mineral grains.

Magnetic anomalies may be related to primary igneous or sedimentary processes that establish the magnetic mineralogy, or they may be related to secondary alteration that either introduces or removes magnetic minerals. In mineral exploration and its geo-environmental considerations, the secondary effects in rocks that host ore deposits associated with hydrothermal systems are important and magnetic surveys may outline zones of fossil hydrothermal activity. Because rock alteration can effect a change in bulk density as well as magnetization, magnetic anomalies, when corrected for magnetization direction, sometimes coincide with gravity anomalies.

Magnetic exploration may directly detect some iron ore deposits (magnetite or banded iron formation), and magnetic methods often are an useful for deducing subsurface lithology and structure that may indirectly aid identification of mineralized rock, patterns of effluent flow, and extent of permissive terranes and (or) favorable tracts for deposits beneath surficial cover. Geo-environmental applications may also include identification of magnetic minerals associated with ore or waste rock from which hazardous materials may be released. Such associations permit the indirect identification of hazardous materials such as those present in many nickel-copper or serpentine hosted asbestos deposits.

Gamma-ray Methods

Gamma-ray methods use scintillometry to identify the presence of the natural radioelements potassium, uranium, and thorium; multi-channel spectrometers can provide measures of individual radioelement abundances. Gamma-ray methods have had wide application in uranium exploration because they provide direct detection. Thorium is generally the most immobile of the three radioelements and has geochemical behavior similar to that of zirconium. Thorium content, like uranium content, tends to increase in felsic rocks and generally increases with alkalinity.

Gamma-ray spectrometry, because it can provide direct quantitative measures of the natural radioelements, provides geo-environmental information concerning radiation dose and radon potential. Because uranium and (or) potassium are commonly enriched in or adjacent to some deposits, their presence may often be used to indirectly assess the potential for release of hazardous materials from ore or waste piles. Where sulfide minerals are present their oxidation accelerates uranium mobilization.

Seismic Methods

Seismic techniques have had relatively limited utilization, due to their relatively high cost and the difficulty of acquiring and interpreting seismic data in strongly faulted and altered igneous terranes, in mineral assessments and exploration at the deposit scale. However, shallow seismic surveys employ less expensive sources and smaller surveys than are typical of regional surveys, and the cost of studying certain geo-environmental problems in the near subsurface may not be prohibitive. Reflection seismic methods provide fine structural detail and refraction methods provide precise estimates of depth to lithologies of differing acoustic impedance. The refraction method has been used in mineral investigations to map low-velocity alluvial deposits such as those that may contain gold, tin, or sand and gravel. Applications in geo-environmental work include studying the structure, thickness, and hydrology of tailings and extent of acid mine drainage around mineral deposits.

Thermal Methods

Two distinct techniques are included under thermal methods: (a) borehole or shallow probe methods for measuring thermal gradient, which is useful itself, and with a knowledge of the thermal conductivity provides a measure of heat flow, and (b) airborne or satellite-based measurements, which can be used to determine the Earth's surface temperature and thermal inertia of surficial materials, of thermal infrared radiation emitted at the Earth's surface. Thermal noise includes topography, variations in thermal conductivity, and intrinsic endothermic and exothermic sources.

Borehole thermal methods have been applied in geothermal exploration, but have seldom been used in mineral exploration. However, this method has potential usefulness in exploration and in geo-environmental investigations. Causes of heat flux anomalies include oxidizing sulfide minerals and high radioelement concentrations.

Conditions that may focus, or disperse, heat flow are hydrologic and topographic influences, as well as anomalous thermal conductivity. In geo-environmental applications, oxidation of sulfide bodies' in-place or on waste piles, if sufficiently rapid, can generate measurable thermal anomalies, which can provide a measure of the amount of metal being released to the environment. Borehole temperatures may also reflect hydrologic and hydrothermal systems that have exploration and geo-environmental consequences. Airborne thermal infrared measurements have applications in geothermal exploration, and may have potential in mineral exploration and in geo-environmental applications whenever ground surface temperature is anomalous due to sulfide oxidation, hydrologic conditions, or heat-flow perturbations due to structure or lithology.

Thermal infrared imaging methods are a specialized branch of more generalized remote sensing techniques. Images obtained in this wavelength range may be used for photo geologic interpretation or, if day and night images are available, to estimate the thermal inertia of the surface. Unconsolidated or glassy materials can be distinguished by their low thermal inertia. In places, thermal infrared images can distinguish areas of anomalous silicification.

Electrical Methods

Electrical methods comprise a multiplicity of separate techniques that employ differing instruments and procedures, have variable exploration depth and lateral resolution, and are known by a large lexicon of names and acronyms describing techniques and their variants. Electrical methods can be described in five classes: (1) direct current resistivity, (2) electromagnetic, (3) mise-a-la-masse, (4) induced polarization, and (5) self-potential. In spite of all the variants, measurements fundamentally are of the Earth's electrical impedance or relate to changes in impedance. Electrical methods have broad application to mineral and geo-environmental problems: they may be used to identify sulfide minerals, are directly applicable to hydrologic investigations, and can be used to identify structures and lithologies.

Direct Current Resistivity Method

Direct current resistivity methods measure Earth resistivity (the inverse of conductivity) using a direct or low frequency alternating current source. Rocks are electrically conductive as consequences of ionic migration in pore space water and more rarely, electronic conduction through metallic luster minerals. Because metallic luster minerals typically do not provide long continuous circuit paths for conduction in the host rock, bulk-rock resistivities are almost always controlled by water content and dissolved ionic species present. High porosity causes low resistivity in water-saturated rocks.

Direct current techniques have application to a variety of mineral exploration and geo-environmental considerations related to various ore deposit types. Massive sulfide deposits are a direct low resistivity target, whereas clay alteration assemblages are an indirect low resistivity target within and around many hydrothermal systems. The wide range of earth material resistivities also makes the method applicable to identification of lithologies and structures that may control mineralization. Acid mine waste, because of high hydrogen ion mobility, provides a more conductive target than solutions containing equivalent concentrations of neutral salts.

Electromagnetic Method

Electromagnetic measurements use alternating magnetic fields to induce measurable current in the Earth. The traditional application of electromagnetic methods in mineral exploration has been in the search for low-resistivity (high-conductivity) massive sulfide deposits. Airborne methods may be used to screen large areas and provide a multitude of targets for ground surveys. Electromagnetic methods, including airborne, are widely used to map lithologic and structural features from which various mineral exploration and geo-environmental inferences are possible.

Mise-a-la-masse Method

The mise-a-la-masse method is a little used technique applied to conductive masses that have large resistivity contrasts with their enclosing host rock. In exploration, application of this method is principally in mapping massive sulfide deposits. This method is useful in geo-environmental investigations of highly conductive targets; it has been applied to identify a contaminant plume emanating from an abandoned mine site.

Self-potential Method

Several possible natural sources generate measurable direct current or quasi-direct current, natural electrical fields or self-potentials. The association of a self-potential anomaly with a sulfide deposit indicates a site of ongoing oxidation and that metals are being mobilized; other self-potential anomalies are due to fluxes of water or heat through the Earth. Geo-environmental applications include searching for zones of oxidation and paths of ground water movement.

Induced Polarization Method

The induced polarization method provides a measure of polarizable minerals (metallic-luster sulfide minerals, clays, and zeolites) within water-bearing pore spaces of rocks. Polarizable minerals, in order to be detected, must present an active surface to pore water. Because induced polarization responses relate to active surface areas within rocks, disseminated sulfide minerals provide a much better target for this method than massive sulfide deposits, although in practice most massive sulfide deposits have significant gangue and have measurable induced polarization. Induced polarization has found its greatest application in exploration for disseminated sulfide ore, where it may detect as little as 0.5 volume percent total metallic luster sulfide minerals. In geo-environmental studies, induced polarization surveys are principally used to identify sulfide minerals, but it may have other applications, such as outlining clay aquitards that can control mine effluent flow.

Remote Sensing Methods

Remote sensing includes methods that utilize images obtained in the ultra-violet, visible, and near infrared bands of the electromagnetic spectrum. Thermal infrared observations, discussed previously under thermal methods, are also part of remote sensing. Remote sensing data are treated in image format, often in digital form, so that they can be processed conveniently. By comparison with known spectral responses of minerals or mineral groups, iron hydroxide minerals, silica, clay alteration, etc., can be defined over broad areas. Remote sensing can be used in geo-environmental studies to map surface alteration patterns and to identify anomalous vegetation patterns in areas related to abnormal metal content in soil.

Other Methods

A number of other geophysical or quasi-geophysical methods have been used, or have potential application, in mineral exploration. Application of these methods in geo-environmental investigations has been limited, but should not be dismissed. Some peripheral techniques that have special uses (as in archeology), whose utilization is not widely known in mineral exploration, that may directly apply to shallow geo-environmental investigations. Examples of such techniques are ground-penetrating radar (used to image the shallow subsurface in electrically resistive rock), the piezoelectric method (used in studies of quartz veins), ultraviolet laser induced fluorescence (the Luminex method, used to identify scheelite, hydrozincite, and other fluorescent minerals), airborne gas sniffing (used in mercury exploration), the Russian CHIM (partial extraction of metals) electro-geochemical sampling technique, and radon sensing.

Applications of Geophysical Methods

Abandoned Mine Workings

Abandoned mine workings may be an environmental hazard because of the possibility of subsidence or collapse. They can also channel contaminated ground water flow, particularly if they are located in sulfide mineral-bearing rock and contain water. Several geophysical strategies enable identification of lost mine workings. In general, the deeper and smaller such workings are, the harder they are to locate using geophysical techniques. However, deep or small workings are less likely to collapse, so it may be less important to locate them. If abandoned workings contain metallic tram rails or ventilation tubes, it may be possible to trace such installations from the surface using a magnetometer or geo-electrical device. If part of the metallic installation is accessible, the mise-a-la-masse method can be used to locate the rest of the installation. These methods only trace the metallic installation, so they will not identify workings where the installation is absent or, in the mise-a-la-masse method, where workings are disconnected from the energized installation segment. Further, the method may not map true voids; the tunnel where the metal was installed may have collapsed.

Magnetometers have been used to trace burned-out coal seams, particularly pyrite-bearing coal. During combustion, pyrite may oxidize to magnetite, which then can be traced using surface or airborne magnetic measurements. If the burned seam is still hot, geo-thermometry and infrared sensing methods may be used.

When the above special conditions are not met, the challenge for geophysical techniques is to find large voids, regardless of whether the voids are manmade. More reliable approaches to this challenge involve microgravity, seismic, ground-penetrating radar, geoelectric surveys, or a combination thereof. Microgravity requires measurements be made at a spacing about equal to the diameter of the workings being sought. Only shallow workings, whose tops are located at a depth less than about one opening-diameter, can be traced using microgravity.

Shallow, high-resolution seismic methods and ground penetrating radar have been used to trace caves. Both methods are applicable to identification of air- or water-filled openings.

Argillic Alteration

Argillic alteration that surrounds some sulfide deposits may be associated with resistivity lows, induced polarization highs, and moderate magnetic lows, whereas silicified rock and quartz lenses associated with sulfide deposits are associated with high resistivity, low magnetization, and increased density. Altered tracts commonly have diagnostic reflectance patterns on remote-sensing multispectral images. In addition, trace elements and alteration products may be absorbed by plants resulting in recognizable reflectance anomalies.

Structures Controlling Contaminated Water Flow

Numerous geologic structures that can focus ground water flow can be identified by various geophysical surveys. Aquitards can be identified by direct current or electromagnetic resistivity studies. Clay lenses are less resistive than sand- and gravel-bearing aquifers, unless contained pore water is ion rich or highly acidic, in which case the aquifers may be less resistive. Reflection seismic and ground penetrating radar may distinguish subtle velocity and resistivity changes related to distributions of aquifers and aquitards in sedimentary environments. Pore fluid with properties that vary laterally and with depth cause some uncertainties in aquitard delineation.

Lithologic boundaries and faults may be associated with porosity contrasts that may concentrate ground water flow. Porosity contrasts are associated with density, resistivity, and acoustic velocity contrasts, and some may have associated gamma-ray contrasts. In most cases, increased porosity lowers density and resistivity, the latter resulting from an increase of conductive pore fluid. In crystalline bedrock, faults and lithologic boundaries commonly have density and magnetic contrasts related to rock composition but ambiguities may result from subtle mineralogical changes in associated rocks.

Bedrock topography, including structural highs and buried channels, that influence ground water flow may contrast with overlying sedimentary and alluvial cover in having higher density, magnetization, and resistivity. These contrasts may be identified using gravity, resistivity, ground-penetrating radar, and seismic refraction and reflection. Anomalies may be small and ambiguities in defining bedrock topography or buried channels include equivalent amplitude anomalies from variable properties of both bedrock and overburden.

Geochemical Methods

A prerequisite for this is sufficient knowledge of the relevant geology and the availability of some pertinent rock samples for geochemical analyses. This quantitative evaluation is based on the fact that insoluble organic matter (kerogen) in sedimentary rocks is partly transformed, under the influence of temperature, into petroleum-like components. The kerogen degradation process is experimentally known to be a first order reaction and can thus be adequately described by the Arrhenius equation.

Types of Geochemical Methods

The geochemical methods generally used are:

- Micro gas survey: The area under investigation is divided into profiles. The interval of the profiles is decided, depending upon the work and generally the distance of 2 kms taken. The laying of the profiles is just the same as is being done in case of seismic survey. The samples of the circulating mud were collected at 5 to 10m interval from exploratory wells. After the samples were collected they were taken to the laboratory for degassing.
- Gas Logging: Gas Logging which is one of the geochemical methods of prospecting for hydrocarbons is a continuous method unlike other geophysical methods. Therefore, on exploratory wells where there may be a danger of blow out, application of gas logging is a boon. It records gas shows which are present in the mud in the form of micro concentrations.
- Hydro chemical surveys: This method analyses formation water properties as it is closely involved in the primary mechanism that causes the accumulation, preservation and destruction of oil and gas fields. Water serves as a vehicle in transporting the hydrocarbons from their source bed to a trap, where they accumulate.
- Organo-Hydro chemical Survey: Natural gases are understood to diffuse into edge waters only a few kilometers from petroleum accumulation. Aquifers overlying oil/gas pools show anomalous concentration of gaseous hydrocarbons. Dissolved organics have significantly high concentration in interstitial waters in source rocks during primary migration and in water expelled during clay mineral digenesis.
- Asphaltenes as frontier molecules in geochemical research: Asphaltenes are emerging as frontier molecules in geochemical research as they provide important clues about nature of source organics, maturation migration and secondary alteration effects. It is a major class of petroleum components that has no analogous counterpart in the in the biological system.

Geochemical Exploration Methods

Geochemical Exploration Methods are based on the assumption that the hydrocarbon found in an oil pool tent to migrate upwards because of their lower density, some of these hydrocarbon molecules may eventually reach the surface. In the proved oil/gas fields, the samples of surface are likely to have a comparatively high percentage of hydrocarbon content. Similarly, higher than average chloride content could be expected around the edges of an oil pool left by the water which has migrated and evaporated.

Geochemical method is still in an experimental stage and requires extremely precise analysis technique. It is interesting for an oil explorer because of its direct approach.

Micro Gas Survey

These surveys are prominently carried out in Russia and adjoining countries.

Method

The area under investigation is divided into profiles. The interval of the profiles is decided, depending upon the work and generally the distance of 2 kms taken. The laying of the profiles is just the same as is being done in case of seismic survey. The samples of the circulating muds were collected at 5 to 10m interval from exploratory wells. After the samples were collected they were taken to the laboratory for degassing. The quantity of sample collected was half a litre. The chromatograph was used for analyzing the different samples. The quantity of different gas is calculated with the help of the formula:

$$Q = \frac{L.K.V.1.88}{100}$$

Where,

L= Length of the peak.

K = Derived value.

V = Volume of gas.

For conducting the micro gas survey, the following precautions are necessary:

- The water used for drilling of the holes should be as pure as possible.
- The bottles in which the samples are collected should be clean and washed with warm water.
- The people conducting the analysis of gas in the laboratory should be so done that they are free from the atmospheric gases.
- Generally the collection of the gas samples should be so done that they are free from the atmospheric gases.

Micro Gas surveys can be used in conjunction with seismic survey to improve the quality of work in a big area and to delineate structures. The greatest utility of Gas Surveys has been proved in the following circumstances:

- Prospecting for pools in stratigraphic and lithological traps.
- For deciding whether a previously discovered structural trap does contain oil or simply 'dry'.

Various forms of gas surveys have been tried in other countries of the world. On the basis of these trials, it has been confirmed that the best and least ambiguous results are obtained in 'Deep gas survey' in which the subsoil atmosphere from depth 2.60 meters at various levels is examined for hydrocarbon content. A more effective version of the same is 'gas logging' undertaken in shallow, structural and deep exploratory wells. Gas surveys on experimental basis should be taken up in one particular area first and then extended to other regions.

Gas Core Surveys

In some of the areas in the western world, gas core surveys were earlier done. The surveys were done by degassing the cores collected during drilling and analyzing gas. Such surveys are now being discontinued.

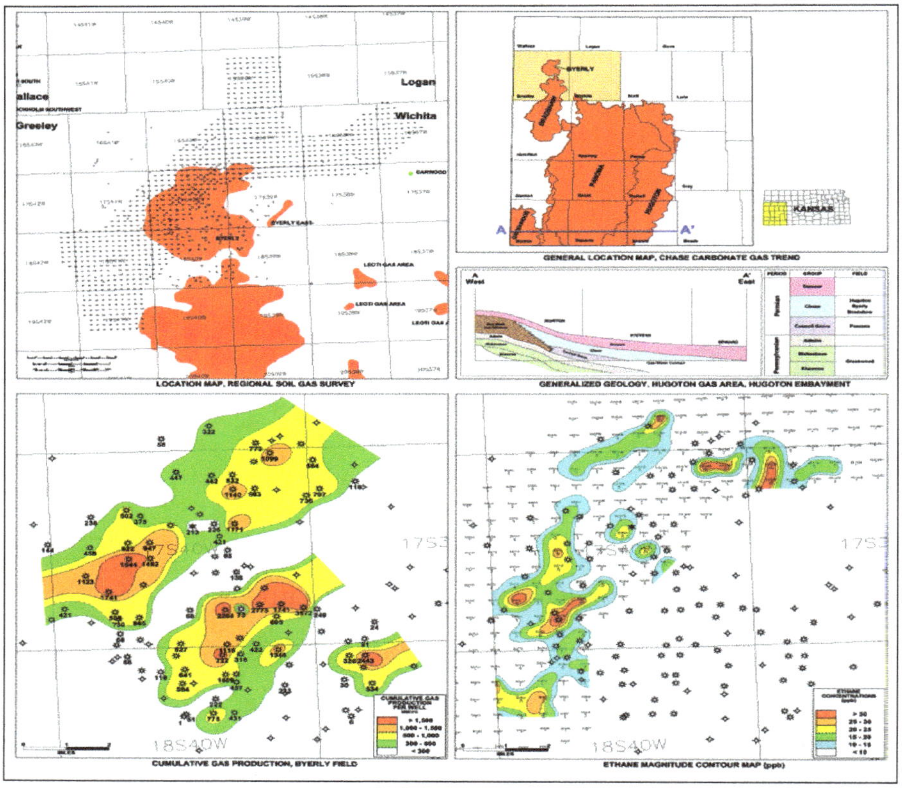

Soil gas survey.

Gas Logging

Gas Logging which is one of the geochemical methods of prospecting for hydrocarbons is a continuous method unlike other geophysical methods. Therefore, on exploratory wells where there may be a danger of blow out, application of gas logging is a boon Much earlier, to the penetration of productive horizons it can prewar about the approaching pool since it has got the advantage of recording the gas shows which area present in the mud in the form of micro concentrations. Much earlier, to penetration of productive horizons it can pre warn about the approaching pool since it has got the advantage of recording the gas shows coming through diffusion process in the mud. Till such time, the mud do not heavily enriched with gas, and its parameters do not change appreciably. Therefore, gas logging is the only means on which can be relied for information of the approaching danger of a blowout. With the latest models of gas logging units, one can have on the spot and at the very moment of drilling, an idea about the quantitative and qualitative nature of the hydrocarbon coming from the pool.

Hydro Chemical Surveys

Another geo chemical method of prospecting for oil/gas is by hydro chemical surveys. As the name suggests, this method analyses formation water properties as it is closely involved in the primary mechanism that causes the accumulation, preservation and destruction of oil and gas fields. Water serves as a vehicle in transporting the hydrocarbons from their source bed to a trap, where they accumulate. Knowledge of the types, class and characteristics of water associated with oil and gas accumulations is needed in geochemical exploration.

Certain constituents dissolved in oil field water are called favorable indicators of hydrocarbon accumulations. Iodide, ammonium salt, organic acid, salts, ethane butane, low sulphate concentrations and the type and class of brine are important. Amount of aromatic hydrocarbon in formation water directly reflects the occurrence of petroleum and can be used to estimate its proximity.

Hitchon and Horn used a statistical technique discremant analysis to show that formation waters are associated with large hydrocarbon accumulations. According to them iodide and magnesium were important discriminators in water from Paleozoic age rocks with that of water from Mesozoic age rocks. As per Suilin's classification the main subdivisions of water will be:

- Sodium sulphate type - Na_2 SO_4.
- Sodium bicarbonate - Na_2 HCO_3.
- Magnesium chloride - $MgCl_2$.

The oil field waters are highly concentrated in chloride, normally sulphated and saturated with calcium sulphate and carbonate. They generally contain more than 1 mg/lit of iodide and 300 mg/lit of bromide with Cl/Br ratios less than 350 and $SO_4 \times 100$ Cl ratios less than 1. Iodide and Bromide are related to bituminous substances and thus to hydro carbon accumulations.

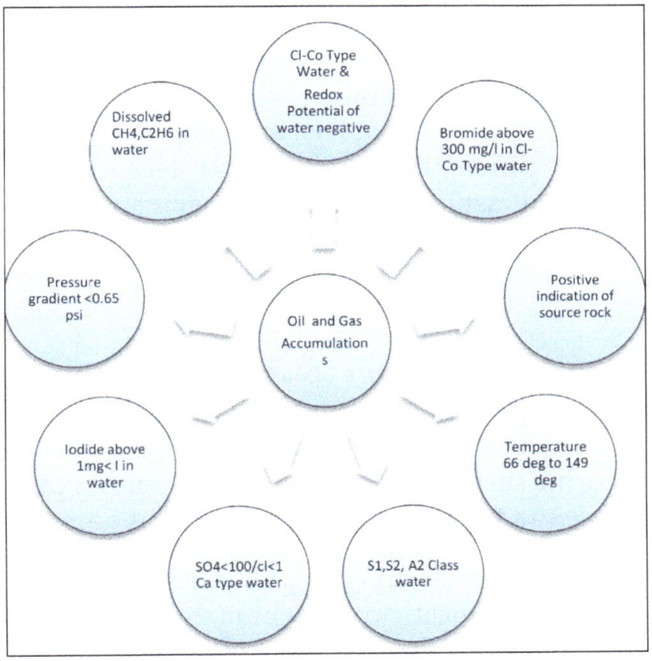

Oil Detection Indicators.

Other important characteristics of oil field water are presence of Benzene (20 mg/lit.), the negative redox potential pressure greater than .65 psi/ft of depth and temperature greater than 66 degree Celsius but less than 149 degree Celsius.

The figure x would show the genetic indicators related to water associated with a reservoir likely to contain oil or gas, while the other one should show the genetic indicators related to water associated with a reservoir not likely to contain accumulations of oil and gas. Appropriate mapping of these indicators with geophysical geological information will prove useful in locating oil and gas pools.

Organo-Hydrochemical Survey

Natural gases are understood to diffuse into edge waters only a few kilometers from petroleum accumulation. Aquifers overlying oil/gas pools show anomalous concentration of gaseous hydrocarbons. Dissolved organics have significantly high concentration in interstitial waters in source rocks during primary migration and in water expelled during clay mineral digenesis. Also aerobic oxidation water washing and microbial degradation leave organo-geochemical imprints.. The most commonly employed hydro chemical indices are:

- Dissolved gas,
- Total dissolved organic carbon,
- Benzene,
- Naphthenic acids,
- Aliphatic acids.

Asphaltenes as Frontier Molecules in Geochemical Research

Asphaltenes are emerging as frontier molecules in geochemical research as they provide important clues about nature of source organics, maturation migration and secondary alteration effects. It is a major class of petroleum components that has no analogous counterpart in the in the biological system. This suggests that petroleum asphaltenes are secondary products formed after the decomposition of source material. Asphaltenes may be fragments of original kerogen from which petroleum is derived and may be expelled as a part of oil.

Sniffer Surveys

The detection of oil and gas seeps in the offshore area by method of sniffing is now being conducted on a routine basis and is one of the most important geochemical methods being followed in the offshore areas. Hydrocarbons seeping from the sea floor dissolve in the sea water and form plumes which are transported by marine currents and mixing. These plumes can be detected at a distance of 10 km from their source area and are sometimes detectable as far as 20 km.

In order to optimize the probability of detecting seep continuous sampling must be conducted at a depth below thermocline. Analytical sensitivity must be of the order of 5×10^{-9} ml gas per ml water. In order to find out the place from where the seep is originating and to simplify data interpretation, the response time of the analytical system must be short. A computer analysis of a mathematical model has been developed to determine optimum lane spacing.

This system provides for direct detection and is specific for hydrocarbons. It would be economical if conducted abroad a vessel on conventional geophysical survey. The sniffer data has been collected and used both in the time of reconnaissance survey and during detailed exploration survey. At the time of reconnaissance, it is done with the conventional seismic equipment. The data obtained with the help of sniffer survey is utilized to confirm that the area of interest is a petroliferous area. Seeps observed are plotted on large scale map and their areal distribution and relationship to subsurface geology and structure are noted.

When this is done along with other geophysical surveys, the sniffer operator makes notes of anomalous areas which should be given special attention during the data interpretation phase. When deployed on a survey where the sniffer is considered to be particularly important the operator make certain operational decisions based on the data at the movement. The sniffer data consists both of hydrographic parameters which effect the distribution of hydrocarbon anomalies and hydrocarbon concentrations. The vessel carrying sniffer survey equipment provides a continuous measurement of the salinity and the temperature and depth sensors located in the towed body. The continuous values of each of the Hydrocarbon components are also plotted. Sniffer surveys are thus a direct help in locating hydrocarbons, where the seismic surveys also indicate promising structures in the offshore areas.

Petroleum Extraction

Because petroleum is a fluid, it is able to migrate through the earth as it forms. To form large, economically recoverable amounts of oil underground, two things are needed: an oil pool and an oil trap. An oil pool, which is the underground reservoir of oil, may literally be a pool or it could be droplets of oil collected in a highly porous rock such as sandstone. An oil trap is a non-porous rock formation that holds the oil pool in place. Obviously, in order to stay in the ground, the fluids – oil and associated gas – must be trapped, so that they cannot flow to the surface of the earth. The hydrocarbons accumulate in reservoir rock, the porous sandstone or limestone. The reservoir rock must have a covering of an impervious rock that will not allow the passage of the hydrocarbon fluids to the surface.

The impervious rock covering the reservoir rocks is called a cap rock. Oil traps consist of hydrocarbon fluids held in porous rock covered by a cap rock. A hot, wet climate fosters the growth of large amounts of organisms. If this growth takes place in a shallow sea, the eventual drying out of the environment and evaporation of the sea water leaves behind large deposits of salt. Salt makes an excellent cap rock for a reservoir. If these conditions are enhanced by a gentle geological folding of the subsurface rocks, the rock folding can produce very large reservoirs, with the impervious salt deposits acting as a cap. These are precisely the conditions that prevailed in the Middle East, giving rise to the enormous deposits of oil found in that region of the world.

Crude oil is extracted by creating pressure gradients within a reservoir that propel the liquid to the well. It is currently recovered in three main stages: primary, secondary, and tertiary recovery.

Primary recovery utilizes the natural energy of the reservoir to drive the hydrocarbons to the surface. Secondary recovery applies an outside agent to drive the hydrocarbons to the surface. Finally, tertiary recovery (also known as Enhanced Oil Recovery) refers to a variety of techniques that are used to improve production that generally involve changing the properties of a reservoir.

Primary Recovery

Primary recovery is the initial method of oil recovery that involves utilizing the natural pressure within the well to extract hydrocarbons with a lift such as a pump or jack. Primary recovery is very limited because it can only target the liquid fraction of the oil, which accounts for between five and fifteen percent of the well's supply. Five drive mechanisms dictate the success of primary recovery. These mechanisms are combination drive, gravity drainage, water drive, gas cap drive, and solution gas drive. The drive mechanisms operate based on several different energy sources within the well.

Before extraction of the hydrocarbon from the reservoir begins, it is nearly impossible to know which drive mechanism is the main contributor. During the early stages of extraction, the production data is analyzed by examining the reservoir pressure and fluid production ratios to determine the primary mechanism. The ratios are revealing because each drive has a unique reservoir pressure.

Solution Gas Drive

A solution gas drive mechanism occurs when a reservoir does not initially contain free gas, but upon depletion of pressure within a reservoir, gas is developed. This drive mechanism requires the reservoir rock to be completely encased in impermeable cap rocks. During extraction, the pressure within the reservoir is lowered. Lowering the pressure causes the source rock, water, and dissolved gasses in the reservoir solution to expand. Energy from the expanding components propels the oil up the collection pipe.

A solution gas reservoir can be considered saturated or unsaturated depending on the existence of free gas within the reservoir. If no free gas exists, then the system is under-saturated and derives energy solely off the expansion of the rock and water. If the system is saturated, then a dissolved gas existing within the reservoir fluid assists in the propulsion of the crude oil. Some form of a solution gas drive can be observed in nearly every other form of primary recovery.

Gas Cap Drive

Some reservoirs contain a natural gaseous divide between the oil column and the cap rock. This phenomenon is referred to as a gas cap. Reservoirs with a gas cap exhibit a gas cap drive mechanism. In his mechanism, the gas pushes down on the oil. The compressed oil is then propelled into the extraction pipes.

Water Drive

In a reservoir fueled by a water drive, an aquifer interacts with the oil and provides the drive energy. The aquifer pushes the oil up the pipe, and as oil is removed, the aquifer continues to rise towards the surface. The success of a water drive reservoir depends on the production rate of the reservoir and the size and permeability of the aquifer. It is more beneficial to have a large, highly permeable aquifer, than a high production rate. This is true because if the oil is slowly extracted the aquifer can rise to replace all space and maintain the pressure. If the aquifer is not permeable enough or too small, it cannot keep up with the rate of flow and the reservoir will lose pressure.

Gravity Drainage

Gravity drainage is a mechanism that works based on density differences between oil, gas, and water. The differing densities promote a natural separation within a reservoir. As the elements flow to their respective zones, a minimal driving force is created. Gravity drainage is largely ineffective on its own but works well when combined with other drive mechanisms.

Combination Drive

It is very uncommon for a reservoir to utilize only one drive mechanism. A typical reservoir will have at least two of the main drive mechanisms, but often more. A combination drive refers to the collect reservoir as a whole and all the coexisting mechanisms that drive its production.

Secondary Recovery

Secondary recovery begins when the natural driving mechanisms have been depleted, therefore requiring human intervention to recover the crude product efficiently. The most common means of improving reservoir efficiency are water flooding and gas injection. These methods involve injecting either water or gas into the base of a reservoir to maintain pressure and to force the oil towards the surface. Secondary recovery produces approximately 20-30% of the oil in a well.

Enhanced Oil Recovery

Current primary and secondary recovery methods are very inefficient. On average, they are only able to recover 33% of the oil in a reservoir. Due to the economics related to the cost of production, most reservoirs are abandoned with around 70% of the product still in the ground.

Enhanced oil recovery surpasses secondary recovery because it injects a specialized solution into the reservoir to maintain pressure as opposed to just water or gas. Even though enhanced oil recovery is better than primary and secondary recovery, it still leaves behind nearly 25% of the oil in a reservoir. Engineers have been experimenting with supercritical carbon dioxide to improve the recovery rate. Under supercritical conditions, carbon dioxide becomes a powerful solvent that readily dissolves in the oil to reduce viscosity. This is notable because the last remaining fractions of oil in a reservoir tend to be thick sludge. By reducing the viscosity, this sludge is able to flow more easily through production pipes.

Oil Well

An oil well is a hole dug into the Earth that serves the purpose of bringing oil or other hydrocarbons - such as natural gas - to the surface. Oil wells almost always produce some natural gas and frequently bring water up with the other petroleum products.

Life Cycle of an Oil Well

Exploration is a method used for searching potentially viable oil and gas sources through geological surveys and drilling exploration wells to identify areas of potential interest. During the drilling process, general information and samples are collected to know about the rocks, fluids to find out how much oil and gas may be available at the explored area and what is the depth of the oil and gas window.

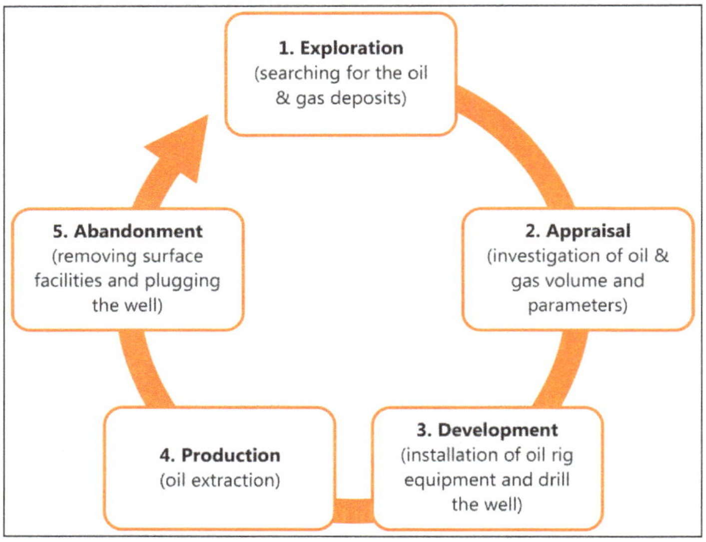

After successful drilling exploration wells, the appraisal stage of the lifecycle starts. The main purpose of this phase is to improve the field description through further data acquisition and to reduce the uncertainty or possibility of losses about the size, shape and marketability of the oil and gas reservoir.

The development stage occurs after successful appraisal and before production. The main activities are formation of a conceptual development plan (in order to develop the oil and gas field, to prepare design for the production wells, to decide what surface and subsurface facilities are required and to describe operating and maintenance principles) and construction of the facilities and production units.

The production phase starts with the first oil flow in the wellhead. Oil and gas fields have a lifespan ranging from 15 to 30 years (from first oil to abandonment) and may be extended up to 50 years or more for the largest deposits. After extraction, oil and gas transported for processing and distribution. When the oil and gas production is no longer cost-effective, wells are plugged and abandoned, production facilities are removed and this is the last stage of oil and gas fields' life cycle.

Types of Oil Well

Sidetrack Wells

A typical operation is to sidetrack a well. This is where a well has already been drilled or partly drilled and there is a need to exit out of one side of the well to a different target. A sidetrack may be required if there is an object stuck in the original hole, which cannot be fished out. In producing fields, an existing well may be sidetracked if there is no further use for that well, e.g., the oil well has watered out. A window will be cut in the casing of the original well by a special milling assembly, and drilling will then proceed out of the window toward a new target.

Horizontal Wells

Horizontal wells are wells where the reservoir section is drilled at a high angle, typically with a trajectory to keep the well within a specific reservoir interval or hydrocarbon zone. In a strict sense, these wells are rarely perfectly horizontal, but they tend to be near horizontal mostly, generally at an angle greater than 80° from vertical.

Horizontal wells are drilled in a specific configuration. The tangent section of the well is drilled along a deviated well path to just above the reservoir section, to what is known as the kick off point. From the kick off point, the well is drilled at an increasingly higher angle, arcing around toward an angle close to horizontal. The point at which the well enters (or lands on) the reservoir is called the entry point. From there on, the well continues at a near-horizontal orientation with the intention of keeping it substantially within the reservoir target until the desired length of horizontal penetration is reached.

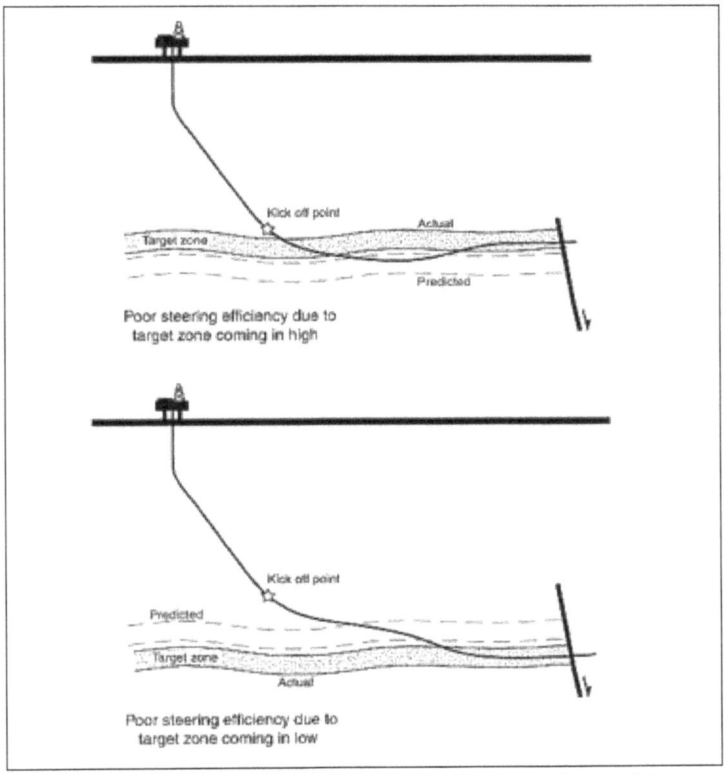

Problems can be encountered with landing a horizontal well if the target zone is too high or too low compared to what is predicted.

One problem in drilling a horizontal well is in locating the kick off point at about the right distance above the reservoir. The kick off point will be planned for a specific depth above the prognosed target zone depth, such that there will be enough room to turn the well around, so as to enter the target at a near horizontal angle. If the target zone comes in high on prediction, the chances are that the well will be drilled all the way through the reservoir before being able to turn round quickly enough to establish a horizontal trajectory. If the target zone is deeper than expected, then quite a long distance of well can be drilled at a very high angle before the reservoir is entered. Given the normal uncertainty on establishing the depth of a target zone from the seismic data, it is common for a pilot hole to be drilled first to get this information directly. Pilot holes may be vertical, although it is better to deviate the pilot-hole in the direction of the horizontal well path, and closer to the planned entry point for the horizontal section. If a horizontal well is planned near an appraisal well, then this can be used as a proxy for a pilot hole.

A horizontal well can be drilled geometrically where there is a reasonable confidence in the expected reservoir geometry. The targets are defined at the entry point and at total depth, and the well is drilled according to a set geometrical plan between them. The alternative is to geo-steer a horizontal well, particularly where there is less confidence in predicting the reservoir geology. Geo-steering involves using geological information obtained as the well is being drilled to try and keep the well path within the target. This can involve the use of real-time log data but may also include input while drilling from well-site biostratigraphy or from examination of drill cuttings if the lithologies at the top and base of the reservoir are distinctive.

The main technique in geo-steering involves the use of a real-time log data display while the horizontal well is being drilled. The down-hole log data can be directly transmitted to a computer screen in the geologist's office from the well site. This allows the geologist to establish which part of the reservoir is being drilled through and then decide where the well should be steered to next. This is done by comparing the real-time logs with data from nearby wells. Log responses in horizontal wells can look different from that in conventional wells. A catalog of expected log responses, as they would appear in a horizontal trajectory, can be created by computer modeling. If the geologist thinks that the well is above the target zone, they will ask the directional driller at the rig site to steer down; if the geologist believes they are below the target, they will ask the driller to steer up.

Designer Wells

Designer wells are types of high-angle or horizontal wells that have more than one intended target. This makes them more cost effective because the individual targets would have otherwise required several conventional wells to drain them effectively. One aim of a designer well could be to penetrate and drain more than one fault block. In mature fields, multi-target infill wells can increase the chances of finding an economic volume of oil. For example, in the Oseberg field, Norwegian North Sea, a designer well successfully targeted and lined up several fluvial channel sandstone bodies.

Multilateral Wells

Multilateral wells are wells that have more than one branch radiating from the main borehole. Each branch can drain a separate part of the reservoir and produce into a common single wellbore. The advantage of multilateral wells is that, for the same number of drainage points, they can be somewhat cheaper than if separate wells had been drilled.

Coiled Tubing Drilling

Coiled tubing is continuous, small-diameter steel pipe stored on a reel at the surface in lengths of up to 6000 m (19,685 ft) long. Coiled tubing can be used in place of drill pipe for new wells and short-length to medium-length horizontal sidetracks (typically with a step-out of less than 800 m [2625 ft]). A mud turbine and drill bit combination is used for coiled tubing drilling. The turbine is powered by the mud moving through it; the tubing itself does not rotate. The advantage of coiled tubing drilling is that the drilling operation is quicker than normal drilling in that the connection time involved with a jointed drill pipe is eliminated. The tubing is simply rolled in and out of the well.

Drilling Techniques

The Drilling Operation

The crew sets up the rig and starts the drilling operations. First, from the starter hole, they drill a surface hole down to a pre-set depth, which is somewhere above where they think the oil trap is located. There are five basic steps to drilling the surface hole:

- Place the drill bit, collar and drill pipe in the hole.
- Attach the Kelly and turntable and begin drilling.
- As drilling progresses, circulate mud through the pipe and out of the bit to float the rock cuttings out of the hole.
- Add new sections (joints) of drill pipes as the hole gets deeper.
- Remove (trip out) the drill pipe, collar and bit when the pre-set depth (anywhere from a few hundred to a couple-thousand feet) is reached.

Drilling continues in stages: They drill, then run and cement new casings, then drill again. When the rock cuttings from the mud reveal the oil sand from the reservoir rock, they may have reached the final depth. At this point, they remove the drilling apparatus from the hole and perform several tests to confirm this finding:

- Well logging: Lowering electrical and gas sensors into the hole to take measurements of the rock formations there.
- Drill-stem testing: Lowering a device into the hole to measure the pressures, this will reveal whether the reservoir rock has been reached.
- Core samples: Taking samples of rock to look for characteristics of reservoir rock.
- Mud logging: It involves the monitoring and recording of a variety of data related to the borehole and to the drilling process. Mud logging incorporates gas analysis and cuttings data with drilling information to build a continuous formation evaluation record as the well is being drilled. Equipment and services for mud logging can range from simple monitoring to integrated computer modeling of the well site and borehole environment.

Equipment

Standard mudiogger's work area. (A) Chromatograph. (B) Chromatograph chart. (C) Back-up chromatograph. (D) Depth and time monitoring equipment. (E) Power panel.

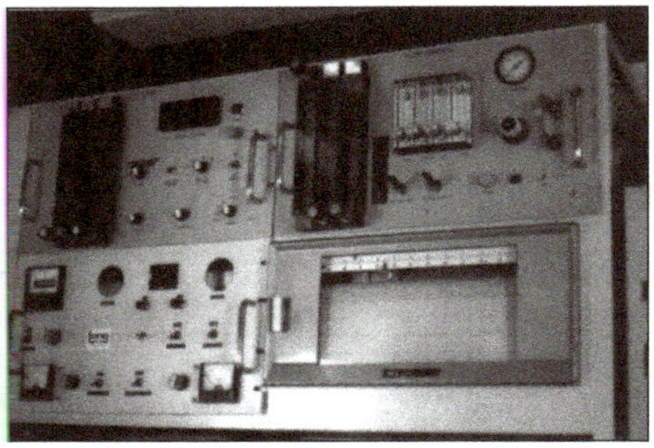

Flame ionization detection (FID) chromatograph.

Standard mud logging unit exterior.

General layout of mud logging unit. (A) Sample examination area. (B) Mudlogger's work area. (C) Sample preparation area.

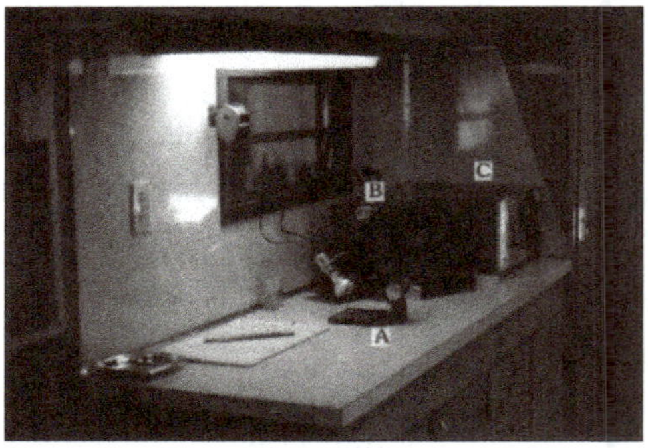

Sample examination area. (A) Binocular microscope. (B) Ultraviolet box. (C) Radioactivity metering device.

Standard Equipment

The minimum complement of equipment for basic mud logging is listed here. Figure above shows much of this equipment in a typical mudiogger's work area.

- Gas extraction: An automatic gas trap is located in the return mud stream for continuous gas extraction.

- Vacuum system: A vacuum pump, pressure regulators, flow meters, filters, and dryers maintain uniform gas sample flow rate and composition from the gas trap and blender. This system includes a pressure and flow alarm and purge system to detect vacuum system blockage or leaks.
- Total gas detector: Hot wire, flame ionization, or infrared detectors are acceptable for total gas detection.
- Chromatograph: An automatic gas chromatograph is equipped with a flame ionization detector (FID) or infrared detector capable of accepting samples from either the gas trap or blender.
- Gas calibration: A gas calibration system injects known single and mixed gases into the total gas detector and chromatograph in a range of concentrations for calibration.
- Blender: A blender agitates and grinds a discrete sample of mud and cuttings and extracts the released gas for measurement.
- Depth and rate of penetration monitor: A sensor independent of the drilling system is capable of detecting changes in depth and compensating for rig motion offshore.
- Pump rate counter: A pump stroke counter is needed for each mud pump.
- Mud pit level monitor: Continuous monitoring of mud pit level(s) and rig motion-compensated volume is necessary.
- Chart recording: A continuous analog chart recording is made of all logged data.
- Mud test kit: This kit contains equipment to perform routine drilling fluid analyses and titrations.
- Sample processing equipment: This equipment is used for catching, sieving, preparing, and testing lithology, mineralogy, and oil content in cuttings.
- Optical equipment: A stereomicroscope is equipped for use with natural and ultraviolet light.
- Hog production equipment: A typewriter, plotter, and copy system are needed for log reproduction.

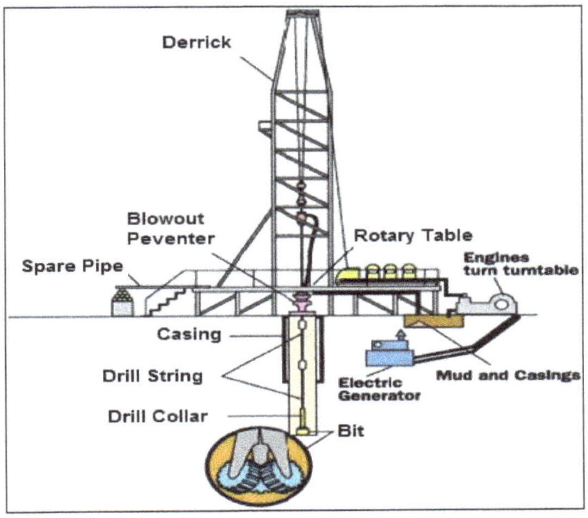

Rotary drilling set up.

Additional equipment may be needed to perform special on-site analyses. Once the well is completed, they must start the flow of oil into the well. For limestone reservoir rock, acid is pumped down the well and out the perforations. The acid dissolves channels in the limestone that lead oil into the well. For sandstone reservoir rock, a specially blended fluid containing proponents (sand, walnut shells, aluminum pellets) is pumped down the well and out the perforations. The pressure from this fluid makes small fractures in the sandstone that allow oil to flow into the well, while the proponents hold these fractures open. Once the oil is flowing, the oil rig is removed from the site and production equipment is set up to extract the oil from the well. Figure below shows the layout of a drilling rig.

The Hole Making Process

It is the process of creating a path in subsoil, includes breaking rock in consolidated sediments and stirring in unconsolidated sediments. Most of the hole making process is through mechanical methods, but other means of high frequency vibration & Hydraulic pressure is also being used. Mechanical means of hole making are classified as follows:

- Cable tool drilling: Heavy string of drilling tools frequently lifted & dropped into hole breaking the consolidate rock into fragments. These fragments are formed into slurry and lifted out of hole by bails.
- Auger Drilling: Done with a helical screw driven into ground with rotation, the drilled earth is lifted by the blade of helical screw. Repeated same till the required depth is used. Used for site investigation, environmental/Geotechnical drilling & sampling, bore holes for construction process.

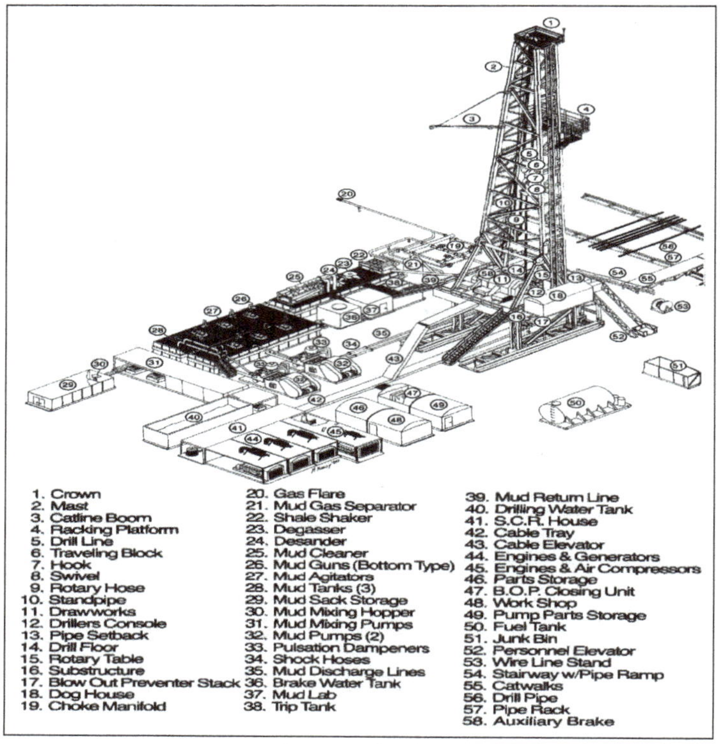

Layout of a rig.

- Rotary Drilling: Rotating sharp drill bits used to dig into earth's crust. Drilled cuttings are picked up & cleared out of the bore hole by drilling fluid.

  In rotary drilling, the disintegration of the rock occurs as a result of a concurrent action on the bit of a load and a torque it shears it. Cuttings from the hole are carried to the surface by drilling mud which is circulated continuously through the drill string. The mud is pumped into the drill pipe at the surface, out through the bit, and up the annular space between the drill pipe and the walls of the hole.

Drilling Phase

During the wellbore making phases, there are several operations performed at the same time to have better control over the process.

Logging Operation

Well logging is a technique to record geophysical properties of rock as a function of depth. It is accomplished by moving a down hole logging probe and recording its sensor output at the surface through an electrical cable. The borehole logging probe or the well logging tool relates to a particular physical property of the rock and mud system. The well logging system consists of:

- Down hole-logging tool.
- Surface computer system.
- Wire line cable for transmitting down hole signal to surface system.

Majority of logs are recorded while pulling the tool upward in the borehole. The logs are recorded both in the open hole as well as in the cased hole, though with different objectives. There is a large variety of logging tools meant for different types of measurements. Generally these tools are named according to what parameter they measure or to the physical process involved during their operation e.g. the tool which measures resistivity of the formation is resistivity log.

S.n.	Name Log	Measured Parameter
1	Caliper log	Bore hole size deviation.
2	Spontaneous	Sand shale boundaries.
3	Gamma ray log	Sand shale boundaries / well to well correlations/depth Correlations.
4	Resistivity log	Measures resistivity / support tool along with other tool.
5	Sonic acoustic log	Porosity.

Coring Operation

Particularly during the exploration phase of a field, coring presents an important means to calibrate the petro-physical model and gain additional information about the reservoir not obtainable by logs. Usually the decision of when and where to core will be made in conjunction with the geologist and operations department, taking into account the costs and data requirements.

Cementing Operation

Oil well cementing is the process of mixing and displacing a slurry down the casing and up the annulus behind the casing, where is allowed to "set", thus bonding the casing to the formation. Some additional functions of cementing includes:

- Protecting producing formations.
- Providing support for the casing.
- Protecting the casing from corrosion.
- Sealing off troublesome zones.
- Protecting the borehole in the event of problems.

The main ingredient in most cement is "Portland" cement, a mixture of limestone and clay. This name comes from the solid mixture resembling the rocks quarried on the Isle of Portland, off the coast of England. There are two types of cementing process:

- Primary cementing.
- Secondary cementing.

The cementation that take place soon after the lowering of casing is called the primary cementation and any cementation job after primary cementation is known as secondary cementation.

Drilling Problems

The most common drilling problems are the creation of doglegs and key seats, hole instability, lost circulation, and excessive bottom hole temperatures.

Doglegs

A dogleg is usually defined as any deviation greater than 3° per 100 ft, and it occurs when a sharp change of direction is taken in the wellbore. Typically, a dogleg is caused by a change in the dip of the formation or by a change in the weight applied to the bit. Severe doglegs can result in stuck casing drill pipe failure, and inability to run casing to total depth. If casing is successfully run through the dogleg, excessive wear on production equipment can occur. The use of properly placed stabilizers, large diameter drill collars, and the proper weight applied to the bit will minimize the formation of doglegs.

Key Seats

Key seats usually form as a result of dog legs. A key seat is formed when a channel or groove is cut in the side of the hole, parallel to the axis of the hole. The drill pipe dragging action through the sharp bend in a dog log creates the groove in the side of the wellbore. Key seats can be prevented by not creating doglegs.

Hole Instability

Hole instability occurs when encountered formations flow, slough, or swell. The most unstable formations are shales and salt beds. Instability can result from the following phenomena:

- Overburden pressure.
- Earth movement forces.
- Pore pressure.
- Water absorption, swelling, or dispersion.

Instability occurs when the relief of overburden pressure exceeds the yield strength of the drilled formation, resulting in flow of the formation (plastic flow). Abnormally high pore pressures can cause blowouts when present in highly permeable formations. If the pressure differential between the wall of the hole and the fluid in the hole is large, the formation may slough off. Structural stresses can also cause hole instability. Problems associated with hole instability include the following:

- Ineffective hole cleaning.
- Stuck pipe.
- Bridges and fill-up.
- Wellbore enlargement.
- Increased mud volume.
- Increased cost.
- Poor cement jobs.
- Logging difficulties.

Hole instability can usually be controlled by use of the proper drilling fluid.

Lost Circulation

Lost circulation is the complete or partial loss of drilling mud into a formation. The loss occurs when the total pressure exerted against the formation exceeds the total pressure of the formation. Formations susceptible to lost circulation have the following characteristics:

- Cavernous and open fissured.
- Very coarse, permeable, and shallow, such as loose gravel.
- Naturally or intrinsically fractured.
- Easily fractured.
- Under pressured or depleted.

Lost circulation results in increased mud expense and may cause subsurface blowouts. The proper drilling fluids and use of lost circulation material will minimize lost circulation.

Bottom Hole Temperatures

Extremely high bottom hole temperatures can occur in deep wellbores or in areas of abnormally high geothermal gradient. These excessively high bottom hole temperatures (greater than 250°C) can cause drilling problems because of the accelerated thickening of water-based drilling fluids. The increase in viscosity and density of the drilling can may result in the following:

- Reduction in penetration rates.
- Lost circulation.
- The well-being swabbed when drill pipe is pulled.
- Stuck tools.

This problem can be mitigated by using oil-based muds.

Drilling Rig

During the first phase of the development of the well, a rotary drilling rig is installed to bore a hole in the ground and reach the oil reservoir. The main rotary drilling rig components are derrick or mast, power and prime movers, hoisting equipment, rotating component, circulating system, tubular and tubular handling equipment and bit.

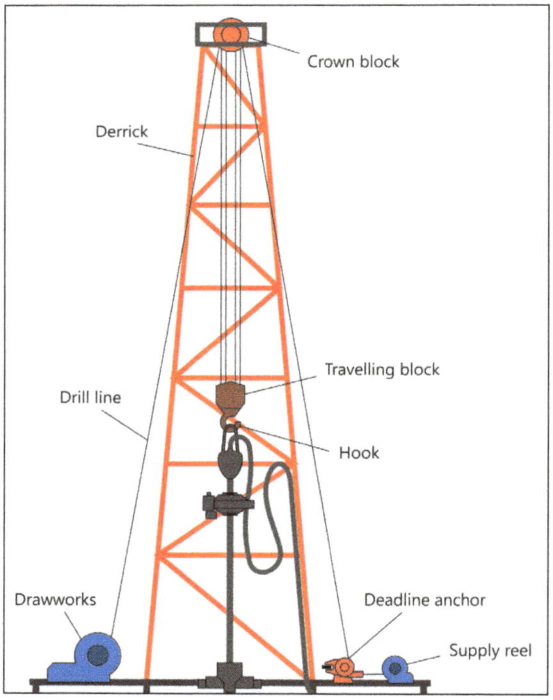

Hoisting equipment of the drilling rig.

- Derrick: Derrick is mainly used offshore and is a large load-bearing vertical structure, usually of bolted construction and pyramidal in shape, for the equipment used to lower and raise the drill string into and out of the wellbore. The height of the derrick does not affect its loadbearing capacity, but it shows the maximum length of the drill pipe section. The standard derrick has square-shaped rig floor with four legs standing at the corners of the substructure. It provides work space for the necessary equipment on the rig floor.

- Mast: Mast is mainly used with onshore rigs and is a portable derrick that can be raised as unit but for the transporting can be divided into two or more sections. It is usually rectangular or trapezoidal in shape.
- Power and prime movers: The power generated by the power system is used for five main operations such as rotating, hosting, drilling fluid circulation, rig lighting and hydraulic systems. It is important to note that the most of the generated power is consumed by the hoisting and drilling fluid circulation systems. Internal combustion engine (mostly diesel) connected to electric generators or turbine is the source of power on the rig. Some rotary rigs can use electricity directly from power lines.
- Hoisting: Hoisting component is used to perform all lifting activities on the rig and helps in lowering or raising equipment into or out of the well. It consists of draw works, crown block, traveling block, deadline anchor, supply reel and drilling line.
- Draw works is the main operating component of the hoisting system and is used to transmit power from prime movers to the hoisting drum that lifts drill string, casing or tubing string out of and to lower it back into the borehole. They consist of a large diameter steel spool, brakes, a power source and assorted auxiliary devices. The primary function of the draw works is to reel out and reel in the drill line, a large diameter wire rope, in a controlled manner. The speeds for hoisting the drill string could be changes by driller via integrated gear system.
- Crown block is fixed assembly of sheaves (single or double) with a wire rope drilling line running between it and is located at the top of the derrick or mast and over which the drilling line is threaded. It is used to change the direction of pull from the draw works to the traveling block.
- Traveling block and hook combination is used to safely and efficiently raise or lower tools and equipment in the well. It is the set of sheaves or pulleys through which the drill line (wire rope) is threaded or reeved, is opposite the crown block and enabling heavy loads to be lifted out of or lowered into the wellbore. Hook is located beneath the traveling block and is used to pick up and secure the swivel and Kelly.
- Deadline anchor is usually bolted on to the substructure and is the equipment that holds down the deadline part of the wire rope. It provides weight measurements and secure deadlines.
- Supply reel is a spool that stores the unused portion of the drill line.
- Drill line is the wire rope used to support the drilling tools. It is threaded or reeved through the traveling block and crown block to facilitate the lowering and lifting of the drill string into and out of the borehole. Drill line then clamped to the rig floor by the deadline anchor.

Rotating Component

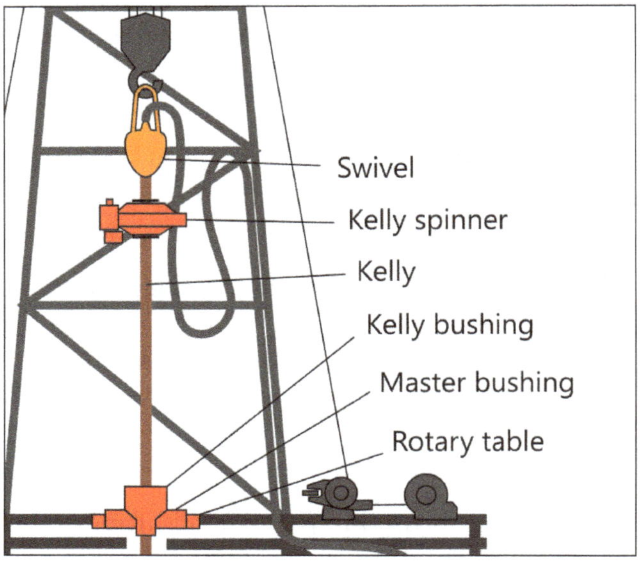

Rotating equipment of the drilling rig.

Rotating component is the equipment responsible for rotating the drill string. It consists of the swivel, Kelly spinner, Kelly or top drive, Kelly bushing, master bushing and rotary table.

- Swivel is a mechanical device that is hung from the hook of the traveling block to support the weight of the drill string and allows it to rotate freely. It provides connection for the rotary hose as well as passageway for the flow of drilling fluid into the drill stem.
- Kelly spinner is a pneumatically controlled device mounted below the swivel to spin the Kelly to make up tool joints when making connections.
- Kelly is the heavy steel square or hexagonal member that is suspended from the swivel through the rotary table and connected to the topmost joint of drill pipe to turn the drill stem as the rotary table turns. It has a hole drilled through the middle that permits fluid to be circulated into the drill stem and up the annulus or vice versa. The Kelly goes through the Kelly bushing, which is driven by the rotary table.
- Top drive is a hydraulically powered device on the drilling rig and is located at the swivel place. It allows the drill stem to spin and facilitate the process of drilling a borehole. Top drive means a power swivel, which directly turns the drill string without need for a Kelly and rotary table.
- Kelly bushing is a device that fits into a part of rotary table called master bushing, transmits torque to the Kelly and simultaneously permits vertical movement of the Kelly to make hole. The Kelly bushing as Kelly is square or hexagonal and has an inside profile matching the Kelly's outside profile with slightly larger dimensions so that the Kelly can freely move up and down inside.
- Master bushing is a tool that fits into the rotary table of a drilling rig to accommodate the slips and drive the Kelly bushing so that the rotating motion of the rotary table can be transmitted to the Kelly.

- Rotary table is section of the drill floor used to turn the drill stem. It has a beveled gear arrangement to create the rotational motion and opening into which bushings are fitted to drive and support the drilling assembly.

Circulating Component

Circulating component is the rig equipment responsible for the movement of drilling fluid within the well as well as solids removal incurred by the drilling fluid. Normally, the circulation would start from the mud pits or tanks that are located besides the rig. Powerful pumps force the drilling through the surface high-pressure connections to a set of valves called pump manifold, located at the derrick floor. Then, the fluid goes up the rig within a pipe called standpipe to approximately 1/3 of the height of the mast. From there, the drilling fluid flows through a flexible high-pressure rotary hose to the top of the drill string. The flexible hose allows the fluid to flow continuously as the drill string moves up and down during normal drilling operations. The fluid enters in the drill string through a special piece of equipment called swivel located at the top of the Kelly. The swivel permits rotating the drill string while the fluid is pumped through the drill string. In wellbore, the drilling fluid then floes down the rotating string and jets out through the nozzles in the drill bit at the bottom of the hole. Drilling fluid carrying the drilled cuttings and flows out the center of the drill bit and is forced back up the outside of the drill pipe between the drill string and walls of the well (annular) onto the surface of the ground where it is cleaned and circulated back to the well. The cleaning process starts from the shale shaker, which is basically a vibrating screen.

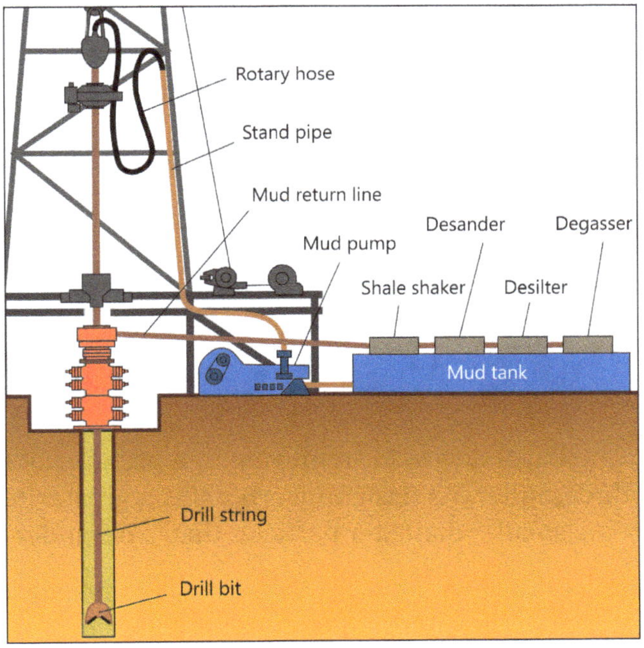

Circulation system of the drilling rig.

This will remove the larger particles, while allowing the residue to pass into settling tanks. The finer material can be removed using other solids removal equipments such as desander and desilter. If the mud contains gas from the formation, it will be passed through a degasser that separates the gas from the liquid mud. Having passed through all the mud processing equipment, the mud is returned to the mud pits or tanks for recycling.

Drilling Fluid Circulation System

The principal components of the drilling fluid circulation system are as follows:

- Mud pump is a large, high-pressure and high-volume pump used to circulate the drilling fluid down the drill pipe and out of the annulus on an oil rig. It could be double acting duplex (2 cylinder) pump, which has four pumping actions per pump cycle or single acting triplex (3 cylinder) with three pumping actions per pump cycle whose pistons or plungers travel in replaceable liners and are driven by a crankshaft actuated by an engine or motor.
- Pump manifold is an arrangement of piping and valves that receives drilling fluid from mud pumps and transmits the drilling fluid to the succeeding circulating component. It is designed to control, distribute and monitor drilling fluid flow.
- Stand pipe is the vertical rigid pipe rising along the side of the derrick or mast, which joins mud pump manifold to the rotary hose.
- Drill string is the mechanical assemblage connecting the rotary drive on the surface to the drilling bit on bottom of the wellbore.

  Mud return line or flow line is the large diameter metal pipe and is the passageway of the drilling fluid as it comes out of the well.
- Shale shaker is the primary solids-removing device with one or more vibrating screens, which is used to remove cuttings from the circulating fluid for reuse. Screens vibrate while the mud flows on top of it. The liquid phase with solids which are smaller than the wire mesh pass through the screen as well as larger solids are retained on the screen and eventually fall to the special container and can be disposed in an environmentally friendly manner.
- Desander is a centrifugal device for removing sand-size particles from the drilling fluid to prevent abrasion of the pumps. There are no moving parts of a desander, and the removal of particles is done by gravity and pressure. As the drilling fluid flows around and gradually down the inside of the cone shape, particles are separated from the liquid by centrifugal forces.
- Desilter is also a centrifugal device for removing free particles of silt from the drilling fluid. Comparing with desander, its design incorporates a greater number of smaller cones, which allow removing smaller diameter particles than a desander does.
- Degasser is device designed to remove air, methane, hydrogen sulfide (H_2S), carbon dioxide (CO_2) and other gases from drilling fluids and allow it to be reused continuously. It helps to reduce the risk of explosions and other dangers during the drilling process.
- Mud pit is an excavated earthen-walled pit and is used only to store used or waste drilling fluid and cuttings.
- Mud tank is an open-top steel container with possibility to observe the consistency of drilling fluid and monitor it level in the tanks. It is used as a reserve store for the drilling fluid.

Tubular and Tubular Handling Equipment

Tubular consists of the following equipments:

- Drill pipe is the longest section of the drill string and is heavy hot-rolled, pierced and seamlessly tubing. It connects the surface equipment with the bottom hole assembly and the bit is used to rotate the bit and for drilling fluid circulation.
- Drill collar is thick-walled, heavy and large diameter steel tube placed between the drill pipe and the bit in the drill stem to provide weight on a bit. It can be cylindrical or spiral shape and is threaded at both ends (male and female) to allow multiple drill collars to be joined above the bit assembly.
- Heavy weight drill pipe is thick-walled tube and is used as transition pipe between drill collar and drill pipe. In high-angled and horizontal wellbore, it is used in lieu of drill collars.
- Subs are short component of the drill string, threaded piece of pipe used to adapt parts of the drilling string that cannot otherwise be screwed together because of difference in thread size or design.

Tubular handling equipment is made of the following equipments:

- Elevator clamps that grip a stand of casing, tubing, drill pipe or drill collars so that the stand or joint can be lifted and lowered into the wellbore opening of the rotary table. The elevators are connected to the traveling block by means of bails, which are solid steel bars with eyes at both sides. Elevator could be side door, center latch or single joint types.
- Elevator links is device designed to support the elevators and attach them to the hook.
- Slips are a wedge-shaped piece of metal with teeth or other gripping elements that supports and transmits the weight of the drill string to the rotary table and are used to hold the pipe in place as well as to prevent pipe from slipping down into borehole. Different types of slips are used during oil well drilling such as drill pipe, drill collar or casing slips.
- Safety clamp is a mechanical device used on tubular above the slips and is used to keep parts of the tool string from falling down the wellbore if other safety measures fail.
- Tongs are large wrenches used to make or break out tubular. It must be used in opposing pairs—make up or breakout tongs to make or break connection.
- Drill pipe spinner is a pneumatically operated device usually suspended on the rig floor used to make fast connections and spin off of drill pipe.
- Iron roughneck is a pneumatically operated machine that replaces the functions performed by the Kelly spinner, drill pipe spinner and tongs and is used to connect and disconnect tubular.

Drill Bits

The drill bit is the tool that attached at end of the drill string that actually cuts up the rock; comes in many shapes and made of materials that are specialized for various drilling tasks and rock formations.

Types of Drill Bit

- Drag/Wing/Blade/Fish tail bits: Having tungsten carbide cutting surfaces for drilling unconsolidated formations. Cutting action is shearing action on the formation.
- Tricone roller bits (TCR bits): Having rollers individually rotate against their respective bearing axis. Cutting action is crushing or chipping. These bits may be of Milled tooth or Tungsten carbide insert bits.
- PDC bits: Polycrystalline diamond compact inserts attached to steel or matrix body blades.
- Diamond bits: Industrial diamonds imparted on them for drilling.

Besides these some special types of bit are available for some special operation. The performance of the drill bit is dependent on many factors such as selection of proper bit, operating parameters and hydraulics. Figure below shows different bit types.

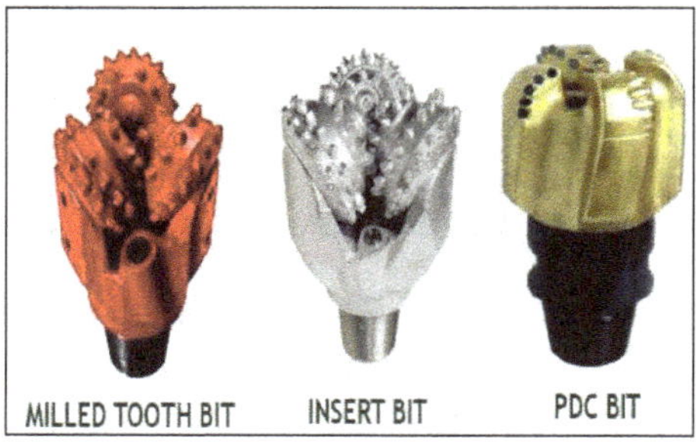

Drill Bits.

Drilling Fluids

Modern drilling fluids (muds) are complex heterogeneous fluids (water based, oil based) and are complex mixtures of more than 200 minerals and chemicals. It is used in a drilling operation and circulates from the surface, down the drill string, through the bit and back to the surface via the annulus. The original use of the drilling fluids was to remove cuttings continuously. Progress in drilling engineering demanded more sophistication from the drilling mud. In order to enhance the usage of drilling fluids, numerous additives were introduced and a simple fluid became a complicated mixture of liquids, solids and chemicals. As the drilling fluids evolved, their design changed to have common characteristic features that aid in safe, economic and satisfactory completion of a well. In addition, drilling fluids are also now required to perform following functions:

- Clean the rock formation beneath the bit for rock cuttings.
- Remove cutting from the well.
- Control formation pressures while drilling and maintain wellbore stability.
- Suspend and release cuttings.
- Seal permeable formations to prevent excessive mud loss.
- Minimize reservoir damage by using reservoir drill-in fluid.
- Cool, lubricate and clean the bit and drilling assembly.

- Transmit hydraulic energy to down hole assembly.
- Ensure adequate formation evaluation.
- Control corrosion.
- Facilitate down hole measurement (measurement while drilling, logging while drilling).
- Facilitate cementing and completion.
- Minimize impact on the environment.

However, excessive use of oil-based drilling fluids may harm the environment and it is important to develop more environmentally friendly drilling fluids. In this respect, water based drilling fluids are more acceptable. As well known, bentonite is widely applied in the water-based drilling fluids, which could enhance the clean properties and form a thin filter with low permeability. The functions of bentonite are to make the fluids more viscous and reduce the loss of fluids.

Water-based Drilling Fluid

Water-based drilling fluid (WBM) is the mud in which water is continuous phase. The water could be fresh, brackish or seawater. The most basic WBM system begins with water, then clays and other chemical and is incorporated into the water to create a homogenous blend. The clay (called "shale" in its rock form or bentonite) is frequently referred to in the oilfield as "gel." Many other chemicals (e.g. potassium formate, $KHCO_2$) are added to a WBM system to achieve various effects, including velocity control, shale stability, enhance drilling rate of penetration, cooling and lubricating of equipment.

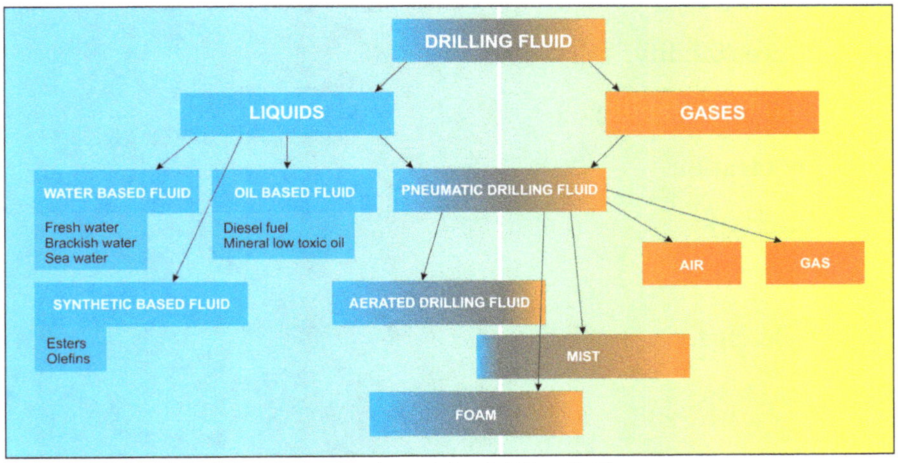

Classification of the drilling fluids.

Advantages of water-based drilling fluid:

- Low cost.
- High rate of penetration.
- Good cuttings removal.
- Good geo-scientific investigations.

- The pressure in the cutting area increases with increasing hydrostatic pressure of drilling fluid.

Disadvantages of water-based drilling fluid:

- Low borehole stability.
- Insufficient cutting transport efficiency.
- Insufficient lubricating properties.
- Drilling fluid loss.

Oil-based Drilling Fluid

Oil-based drilling fluid has best technical properties such as stability, lubricity and temperature stability. Oil-based mud can be a mud where the base fluid is a petroleum product such as diesel fuel or mineral low toxic oil. The authorities do not permit the discharge of oil-based drilling fluid and cuttings drilled with oil-based drilling fluids because of their special nature of being a mixture of two immiscible liquids (oil and water). In that case special treatment and testing are required. The terms oil-based mud and inverted or invert-emulsion mud used to distinguish among the different types of oil-based drilling fluids. Traditionally, an oil-based mud is a fluid with 0–5% by volume of water, whereas an invert-emulsion mud contains more than 5% by volume of water.

Advantages of oil-based drilling fluid:

- Excellent lubricating properties (reduce drilling torque and drag).
- Good temperature stability.
- Favorable to borehole stability.
- High rate of penetration.
- Will not hydrate clays.
- Long bit life.
- Low reservoir damage.
- Low drilling fluid loss.
- Salt not dissolved.
- Corrosion resistance.
- Can be reused.

Disadvantages of oil-based drilling fluid:

- High initial cost.
- Electric log difficulty.

- Viscosity varies with temperature.
- Environmental issue.
- Difficult to keep the rig cleans while drilling.
- Difficult to identify gas kick.
- Messy working environment.
- Fire hazards.

Synthetic Drilling Fluids

These are based on ether, ester or olefin. They have technical properties that are similar to oil-based drilling fluids and are most often used on offshore rigs or in environmentally sensitive areas, because it has the properties of an oil-based mud, but the toxicity of the fluid fumes is much less than an oil-based fluid. This is often used on offshore rigs.

Advantages of synthetic drilling fluids:

- Favorable to borehole stability.
- High rate of penetration.
- Good wellbore stability.
- Good control of drilling fluid properties.
- Good cutting transport efficiency and removal.
- Good filtration properties.

Disadvantages of synthetic drilling fluids:

- Complex system with high solid content.
- Geo-scientific investigations difficulty.

Pneumatic Drilling Fluids

Fluids, which are based on air/gas, mist, aerated fluid or foam are Pneumatic drilling fluids. Air drilling is used primarily in hard rock areas and in special cases to prevent formation damage while drilling into production zones or to circumvent severe lost-circulation problems. Air drilling includes dry air drilling, mist or foam drilling and aerated mud drilling. In dry air drilling, dry air/ gas is injected into the standpipe at a volume and rate sufficient to achieve the annular velocities needed to clean the hole of cuttings. Mist drilling is used when water or oil sands are encountered that produce more fluid than can be dried up using dry air drilling. A mixture of foaming agent and water injected into the air stream, producing foam that separates the cuttings and helps remove fluid from the borehole. In aerated mud drilling, both mud and air pumped into the standpipe at the same time. Aerated mud is used when it is impossible to drill with air alone because of water sands and/ or lost-circulation situations.

Advantages of pneumatic drilling fluids:

- High rate of penetration.
- Low reservoir damage.
- Good bit performance.
- Low drilling fluid loss.
- Low water consumption.
- Low air quality requirements for foam drilling.
- Low hydrostatic pressure.
- Good cleaning of the borehole.

Disadvantages of pneumatic drilling fluids:

- There are restrictions on the possible lithological structures.
- Drilling could be limited by the length of the horizontal section of the well.
- Possibility of fire.
- Possible additional costs to rent equipment.
- Gas costs.
- Gas and foam utilization issues.
- Aerated fluids require specialized equipment for the injections.
- Aerated fluids and foam have potential corrosion problems and the need to use additional inhibitor.
- The quality of the foam changes in exchange pressure.
- The foam is a complicated system and may require computer modeling of foam movement in the borehole.

References

- Petroleum-exploration, upstream: oil-gasportal.com, Retrieved 17, March 2020
- Geochemical-methods-of-petroleum-exploration: ijser.org, Retrieved 14, February 2020
- Cannon-Senior-Essay: yale.edu, Retrieved 25, June 2020
- Petroleum-Extraction-Engineering-322987413: researchgate.net, Retrieved 04, January 2020
- Oil-Well-Rig-and-Directional-Drilling-Technology-A-Field-Investigation-303724192: researchgate.net, Retrieved 20, April 2020

3

Petroleum Reservoirs and Reservoir Engineering

The pool of hydrocarbons present in permeable rocks under the surface of Earth is known as petroleum reservoir. The branch of petroleum engineering which involves applied reservoir mechanics for the development of oil and gas reservoirs is known as reservoir engineering. This chapter has been carefully written to provide an easy understanding of the varied facets of petroleum reservoir and petroleum engineering.

Petroleum Reservoir

Oil and gas accumulations occur in underground traps formed by structural and/or stratigraphic features. Figure below is a schematic representation of a stratigraphic trap. Fortunately, the hydrocarbon accumulations usually occur in the more porous and permeable portion of beds, which are mainly sands, sandstones, limestones, and dolomites; in the intergranular openings; or in pore spaces caused by joints, fractures, and solution activity. A reservoir is that portion of the trapped formation that contains oil and/or gas as a single hydraulically connected system. In some cases the entire trap is filled with oil or gas, and in these instances the trap and the reservoir are the same. Often the hydrocarbon reservoir is hydraulically connected to a volume of water-bearing rock called an aquifer. Many reservoirs are located in large sedimentary basins and share a common aquifer. When this occurs, the production of fluid from one reservoir will cause the pressure to decline in other reservoirs by fluid communication through the aquifer.

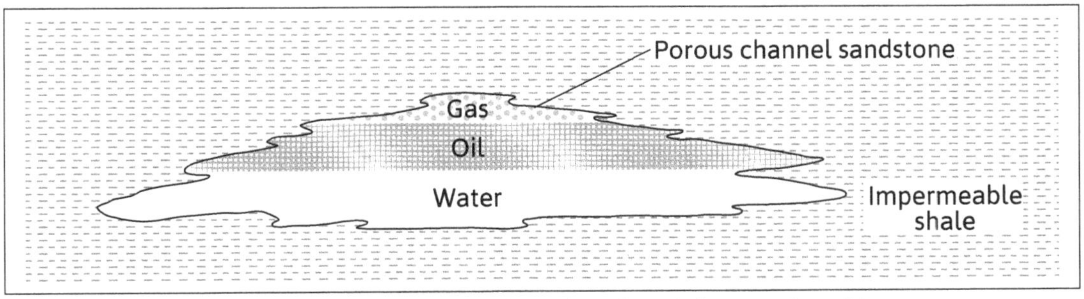

Schematic representation of a hydrocarbon deposit in a stratigraphic trap.

Hydrocarbon fluids are mixtures of molecules containing carbon and hydrogen. Under initial reservoir conditions, the hydrocarbon fluids are in either a single-phase or a two-phase state. A single-phase reservoir fluid may be in a liquid phase (oil) or a gas phase (natural gas). In either case, when produced to the surface, most hydrocarbon fluids will separate into gas and liquid phases. Gas produced at the surface from a fluid that is liquid in the reservoir is called dissolved gas. Therefore, a volume of reservoir oil will produce both oil and the associated dissolved gas at the surface, and both dissolved natural gas and crude oil volumes must be estimated. On the other hand, liquid produced at the surface from a fluid that is gas in the reservoir is called gas condensate because the liquid condenses from the gas phase. An older name for gas condensate is gas distillate. In this case, a volume of reservoir gas will produce both natural gas and condensate at the surface, and both gas and condensate volumes must be estimated. Where the hydrocarbon accumulation is in a two-phase state, the overlying vapor phase is called the gas cap and the underlying liquid phase is called the oil zone. There will be four types of hydrocarbon volumes to be estimated when this occurs: the free gas or associated gas, the dissolved gas, the oil in the oil zone, and the recoverable natural gas liquid (condensate) from the gas cap.

Reservoir Engineering

Reservoir engineering is defined as "the application of scientific principles to the drainage problems arises during the development and production of oil and gas reservoirs"; or "the art of developing and producing oil and/or gas from a reservoir in such a manner as to obtain a high economic recovery".

The main task of reservoir engineering is the optimization of the oil and gas extraction process. To do this, it is necessary to analyze/interpret information generated from useful subsurface measurements, whose objective is the study and characterization of the behaviour of the reservoir. This is accomplished by models, which generate reliable and useful information to evaluate and determine the best exploitation strategy.

Reservoir engineering is also necessary to prevent useless work and expenses (incorrect location or number of wells), initiate control operations, determine whether artificial lift would be necessary, or if secondary recovery or EOR will be required and the best time to implement, always considering the reservoir characteristics and the economic factors.

Technological advancement has driven the development of reservoir engineering. More powerful tools for the engineer are generated with computers of higher speed and greater capacity. This tools and resources are used by the reservoir engineer in their task of optimizing the oil and gas extraction processes.

Within the most representative reservoir engineer tasks are the determination of the STOIIP, the study of the properties of the rock and its relationship with the fluids it contains, the study and analysis of the changes in the physical properties that fluids in the reservoir experience with pressure and temperature changes, the prediction of the behaviour of reservoirs in primary phases of production, determine the pressure variations with time, perform EOR studies and collaborate with work teams towards common objectives related to optimization.

Maturation of Kerogen

Maturation is a combination of complex processes through which complex biological molecules, created by living organisms, are first converted into simple bio-molecule and finally cooked to give petroleum. In the early stages of alteration, termed as diagenesis, an intermediate form of organic matter, called kerogen, is formed. Dehydration (release of H_2O) and Decarboxylation (release of CO_2) are the two main reactions at this stage, dominantly under the influence of bacterial action. The Eh - PH condition at the sediment-water interface principally guides bacterial action on organic matters. While free sulfur is released from sulfate ion with removal of oxygen under the action of Desulfovibrio bacteria in reduced condition below water table, oxydation of free sulfur by the bacteria Thiobacillus in an oxygenated condition result in formation of sulfate ions.

In general, bacterial action at first oxidise organic matters and release water, carbon dioxide, nitrates and phosphates. Reduction reactions follow, which turn nitrates into nitrites and sulfates into hydrogen sulfide and ammonia. Organic matters are diverse and complex in constitution and made up of proteins, carbohydrates, lipids and lignin with proteins being most unstable and lignin most stable. Microbial enzymes decompose these compounds and form bio monomers e.g. proteins transform to amino acids, carbohydrates to sugar and lipids to glycerol and fatty acid. Cellulose gets converted to methane and CO_2. In fact, methane (CH_4) constitutes a major product of bacterial decay. All these changes happen in top few meters of the sediment column and further with increase in depth temperature increases and chemical reactions start dominating over bacterial action.

Net outcome of all diagenetic reactions is the reduction of oxygen content within organic matters keeping carbon: hydrogen ratio intact. As burial continues in deep subsurface at catagenetic stage, fluid phase is released from kerogen (oil and then gas), which further decrease the H:C ratio in the kerogen. A fluid phase high in hydrogen is released and a residue rich in carbon results. In still higher temperature condition at the metagenetic stage, almost near metamorphism, the last hydrocarbon i.e. methane is released leaving behind extremely carbon-rich graphite residue. Although this represents the general pathway through which organic matter and kerogen matures, examples are available where thermally immature Cambrian source beds (about 500 million years old) are exposed in outcrop in southern Sweden, and gas-generating, over-mature Tertiary source beds belonging to Miocene shales encountered along the west flank of the Red Sea in Ethiopia. While in the first case the time elapsed for the Cambrian shales provide the required compensation for the thermal effect, the high thermal gradient in the southern Red Sea, similar to the high gradients found today, is identified as the causative reason in the second case.

The hypotheses about thermal maturation histories of potential source beds must be used carefully. In geochemical modeling of hydrocarbon source rock, geologists widen their range of input data and use data from the fields of coal petrography, such as vitrinite reflection data, or coloration of spores or chitinous microfossils, if they are available. Detailed chemical analyses of oils and source beds for specific compounds also add to our understanding of the thermal histories of rocks and basins. A reliable interpretation of source bed maturation is possible only after all data are collected and processed.

Organic Matters and Kerogen Types

Depending on terrestrial (plant) or aquatic (plant and animal) origin, organic matters are broadly classified as humic and sapropelic in character. Humic organic matters are largely carbohydrate-lignin in composition with protein in subordinate amount, relatively rich in carbon and nitrogen and poor in hydrogen. Thermal alteration of humic organic matter results in humic acids, rich in oxygen and poor in hydrogen. These organic matters are gas-prone and poor in liquid, volatile product yield. In contrast, sapropelic organic matters are of aquatic (marine and lacustrine) derivation, made up of structureless vegetable slime, polymerized lipid and protein. Spores, pollens, cuticles, alginites etc. with high hydrogen content constitute this organic matter and thermal alternation release liquid, volatile products e.g. oil from these organic matters.

Differentiation between the two types of organic matters can be done taking into consideration their H:C contents: H:C content in humic organic matters is less than 0.8, in mixed humic and sapropelic organic matter the ratio is between 0.8 and 1.0 and in sapropelic organic matters it is above 1.0. At diagenetic stage organic matters transform under bacterial attack into a thermally reactive organic complex i.e kerogen with release of methane, CO_2 and H_2O.

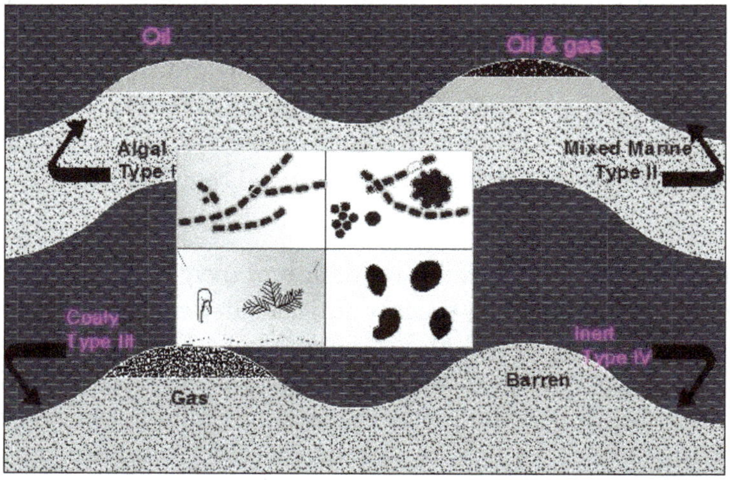

Different types of organic matters; Algal (Type I), Spores, pollens mixed marine (Type II), terrestrial woody (Type III) and Inert, barren (Type IV).

Microscopically, kerogen can be seen as yellow-orange to brown-black particles or amorphous material. Since kerogens originate from different kinds of living organisms with different kinds and proportions of biological molecules, their chemical compositions vary widely and result in formation of hydrocarbons of different types and amounts. Geologists have found it convenient to group kerogens into four fundamentally different classes ranging between Type I and Type IV.

Type I kerogen is derived from sapropelic organic matters, principally constituted of high-grade algal remains, lacustrine, with high H:C ratio i.e. 1.6 - 1.8 and on maturation yields crude oil. It represents the most petroleum generating kerogen type. Type II kerogen consists mostly of amorphous material, both continental and aquatic in origin and derived from the bacterial and mechanical breakdown of a mixture of marine, one-celled plants and animals. Algal tissues, spores, pollens also contribute. This kerogen type has H:C ratio about 1.4 and oil-prone but yields more natural gas than Type I. Type III kerogen (H:C ratio 1.0 or less) is derived from humic organic matters of terrestrial, woody origin from higher land plants and sometimes known as coaly kerogen, equivalent to vitrinite macerals in coal. The humic material in Type III kerogen has a low capacity to form oil and yields mostly natural gas. Type IV kerogen (H:C ratio 0.4 or less) consists mostly of inert particles that have been highly oxidized, recycled and altered before burial, like charcoal. It is the rarest kerogen type and has practically no ability to generate either oil or gas.

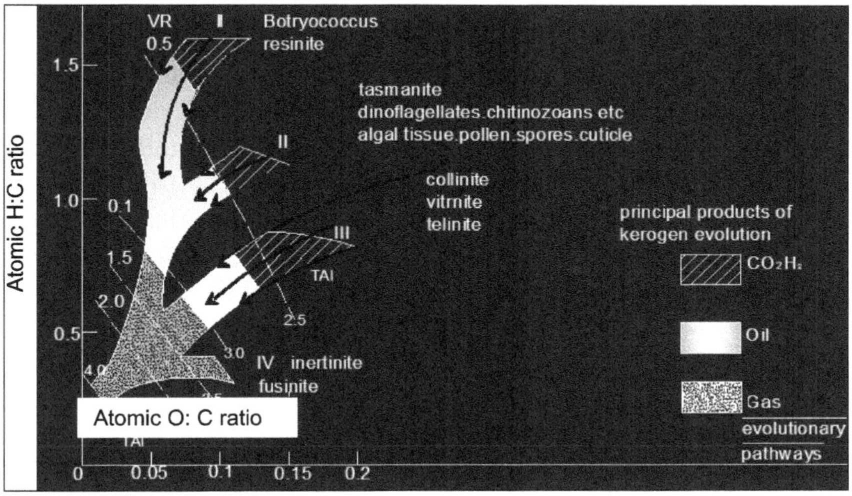

Evolution pathway of different kerogen types with increase in temperature. Dashed while lines represent maturation indices at different stages of transformation. Values on the left of dashed lines represent Vitrinite reflectance (VR) and to the right Thermal Alteration Index (TAI).

The chemistry of crude oil can also be linked to kerogen type and original organic matter. Usually land-derived, non-marine organic matter deposited near continental drainage areas (Type III coaly kerogen) will form mostly gas, but any oil formed will be low sulfur, paraffinic to paraffinic-naphthenic crude oils. Marine organic matter, particularly protein-rich types derived from marine animals (Type I or Type II mixed marine kerogen) tends to yield high sulfur aromatic-intermediate crude.

Biological transformation of organic matter to kerogen and chemical transformation of thermally reactive kerogen give rise to the generation of petroleum. Bacterial processes remain mostly effective in initial brake down of organic matters, while cooking of transformed organic matter i.e kerogen is guided by chemical processes. A combination of negative redox potential and alkaline Ph condition i.e between 7.5 and 8.4 favours liberation of petroleum. In isolated bottom condition where generation of H_2S turns bottom condition acidic with Ph lower than 7, hydrocarbons are retained in kerogen and not liberated. Both temperature and pressure increases as organic matters get buried.

Transformation of organic matters is independent of pressure, in fact high pressure may counter rate of decomposition. In contrast, temperature plays pivotal role in organic matter transformation; higher molecular weight organic matters transform into simpler compounds by thermal cracking. Generation of petroleum is a rate-controlled, thermocatalytic process. In addition to temperature, time spent by organic matter at different temperatures also plays a key role. Since thermal decomposition of organic matter is a temperature-dependant chemical reaction, the rate of transformation reaction is guided by Arrhenius equation i.e $k = Ae^{-Ea}/RT$; where k is rate constant, A is pre-exponential factor dependant on temperature, E_a is activation energy of reaction, R is universal gas constant and T is temperature in Kelvin. According to this equation, reaction rate doubles in every 10 °C increase in temperature. Since generation of petroleum is a chemically reactive process, the amount of petroleum generated will vary linearly with time and exponentially with temperature.

Temperature and Time

Petroleum is generated when kerogen is subjected to the increased temperatures that accompany sediment burial. The alteration of kerogen to petroleum is similar to other thermal cracking reactions. Large kerogen molecules decompose upon heating, to yield smaller molecules of petroleum These reactions usually require temperatures greater than 60 °C. At lower temperatures, during early diagenesis, natural gas, (called biogenic methane or marsh gas) is generated through the action of microorganisms that live near the earth's surface. Vast quantities of biogenic methane are probably generated, but most of this will not encounter a trap and will be lost to the atmosphere.

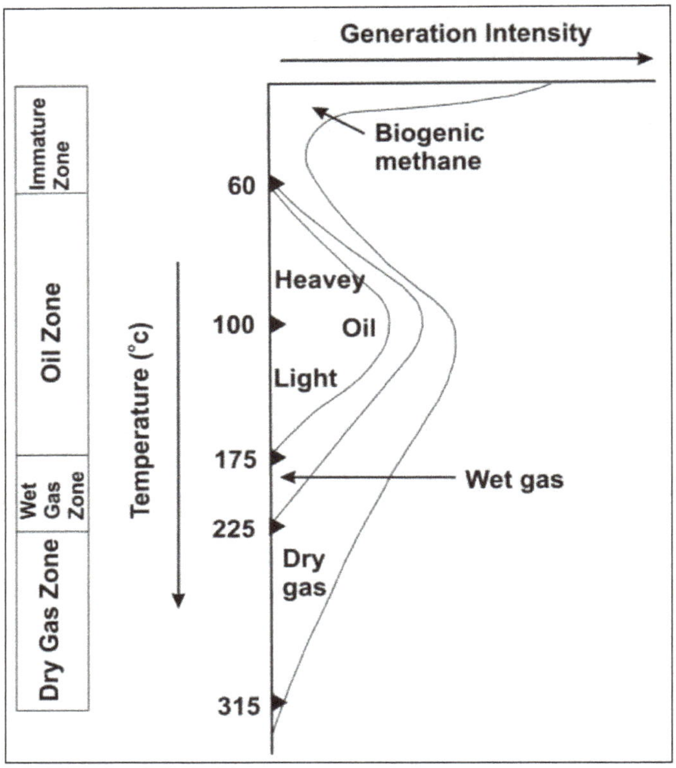

Various transformation products with increase in temperature and burial.
Principle oil producing zone is termed as 'Oil Window'.

The temperature range between about 60 °C and 175 °C is commonly called the oil window. This is the principle zone of oil formation. It begins at burial depths of 1 to 2 km and ends at depths of 3 to 4 km in most areas, depending on factors such as the geothermal gradient. The first oil generated is heavy and tends to be richest in aromatic and N-S-O compounds. As burial and temperature increases, oil becomes lighter and more paraffinic. At temperatures much above 175 °C, the generation of liquid petroleum ceases and gas formation becomes dominant. When formation temperatures exceeds 225 °C, most kerogen has used up its petroleum-generating capacity. Source rocks become overmature. However, some methane can still be created, even at these very high temperatures, by the breakdown of the larger, heavier molecules of previously generated crude oil.

Along with a series of chemical reactions, time also play a major role in the conversion of kerogen to petroleum. Young, Tertiary-age rocks need to be deeply buried or have high geothermal gradients in order to generate significant amounts of petroleum. Although generation, migration and entrapment have been documented in rocks as young as 1.0 to 1.5 million years old, major petroleum accumulations have not been found in rocks younger than 10 million years. On the other hand, some old Paleozoic and Mesozoic source rocks might not have been buried very deeply i.e. buried only up to the uppermost level of the oil window but have still generated petroleum because of the time factor. However, in most petroleum occurrences, temperature appears to be a more significant factor than time.

Petroleum Formation

Most of the oil and gas deposits originate from thick layer of black clay that currently lies several thousand meters under the seabed. The black clay is a source rock, which means a deposit containing significant quantities of organic residue. The clay was deposited around 150 million years ago at the bottom of a sea that covered much of earth's surface. Much of the seabed here was dead and stagnant, while the upper water layers were teeming with life.

As the microscopic phytoplankton died, they sank to the bottom and accumulated in large quantities in the oxygen-free sediments. Over time, they were buried deeper and subjected to a long process of chemical conversion by bacterial decomposition and maturing under a thickening pile of sediment. This caused the formation of liquid and gaseous hydrocarbons in the source rock.

One of the products of anaerobic decomposition of organic matter is kerogen, which at high temperature and pressure slowly generates oil and gas. On the Norwegian continental shelf, the temperature rises by about 25 °C per kilometer of depth. After more than a hundred million years of erosion and sedimentation, the source rock may be buried under several kilometers of clay and sand deposits. Oil is generated when the kerogen temperature reaches 60-120 °C; at higher temperatures, it is mainly gas that is generated.

As oil and gas form, they seep out of the source rock. Because hydrocarbons are lighter than water, the oil and gas migrate upwards in porous water-bearing rock. Oil and gas migration takes thousands of years, and may extend over tens of kilometers until it is stopped by impermeable layers of rock, or the oil or gas leaks out into the sea.

Reservoir rocks are porous and always saturated with water, oil and gas in various combinations. Most of Norway's petroleum resources are trapped in reservoir rocks deposited in large deltas formed by rivers that ran into the sea during the Jurassic Period.

The main reservoirs of the Gullfaks, Oseberg and Statfjord fields are in the large Brent delta that formed in the Jurassic. There are also large reservoirs in sand that was deposited on alluvial plains during the Triassic Period (the Snorre field), in shallow seas in the Late Jurassic (the Troll field) and as subsea fans during the Paleogene Period (the Balder field). In the southern part of the North Sea, thick layers of chalk composed of microscopic calcareous skeletons of plants and animals form an important reservoir rock, as in the Ekofisk Field.

Mudrocks and other impermeable deposits influence migration routes from the source rock to the reservoir. In addition, impermeable rock has to be present to stop petroleum escaping from reservoir rock. Impermeable rock that forms a seal over reservoir rocks is called cap rock. In addition, the configuration of the reservoir rocks must be such that the oil collects in a trap.

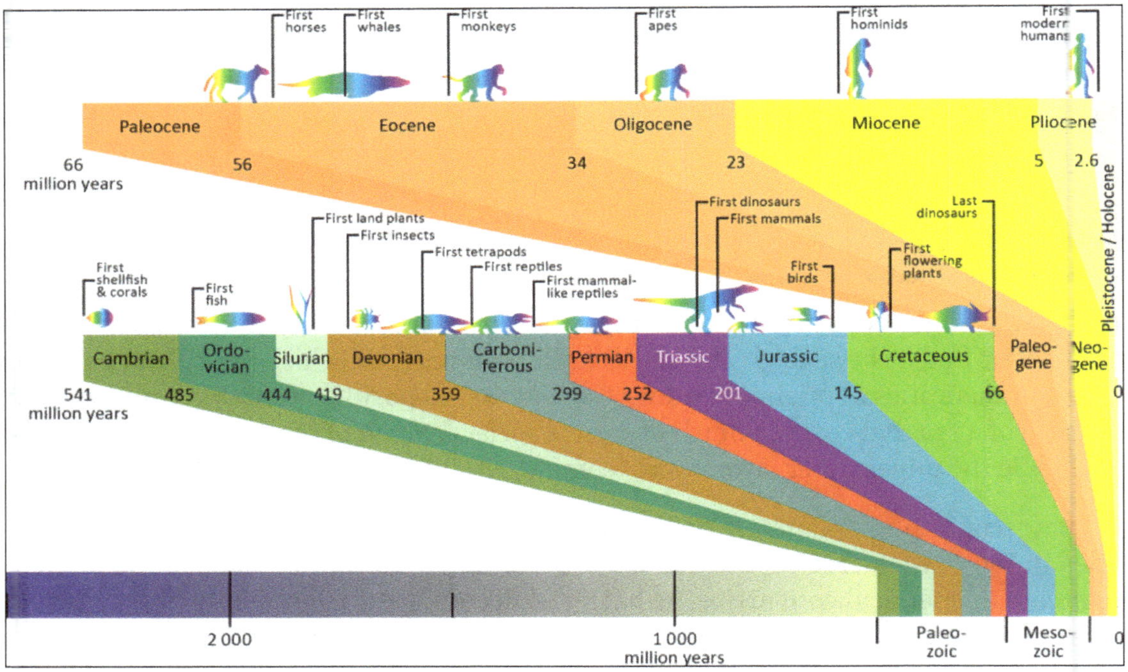

The geological timeline.

If there is a suitable combination of source rock, reservoir rock, cap rock and a trap in an area, recoverable oil and gas deposits may be discovered there.

Petroleum Source Rock

Petroleum source rocks must contain sufficient quantities of sedimentary organic matter with a requisite ratio of hydrogen to carbon. In thermally immature rocks, sedimentary organic matter is dominated by kerogen (organic matter insoluble in common organic solvents) with small amounts of bitumen. Bitumen is organic matter soluble in organic solvents, and the term is used here in a restricted sense for oil generated in petroleum source rocks before it is expelled.

Living organisms and their metabolic processes form sedimentary organic matter. Durand recognized that organic matter decomposes in the water column and in sediments, reducing the quantity and quality (hydrogen-richness) of the organic matter preserved in rocks. Organisms attack sedimentary organic matter for its carbon and hydrogen and metabolically convert some of it to simple molecules (e.g., CO_2, H_2O, CH_4, NH_3, N_2, and H_2S). Abiological oxidation also forms CO_2 and H_2O. These compounds are nonhydrocarbons (except for methane, CH_4) and often escape early in the depositional and burial process. Therefore, the organic residue preserved in rocks and available for thermal conversion to fossil fuels represents only a small proportion of the original biological input. An even lower percentage of this buried fraction is converted to petroleum (oil and gas), expelled, migrated, concentrated, and trapped in reservoirs.

Petroleum source rocks are evaluated by their carbon richness as total organic carbon (TOC) in weight percent (wt. %). They are also evaluated by their hydrogen richness, or quality, as atomic

H/C or, using Rock-Eval pyrolysis, as hydrogen index (HI), which is the ratio of pyrolyzable hydrocarbon mass to rock mass. Thermally immature hydrogen-rich organic matter ($H/C > 1$ or $HI > 300$) has oil-generative potential, whereas lower hydrogen content usually denotes gas-generative potential.

Heat transforms thermally immature sedimentary organic matter to oil and gas. This thermal maturation process is a function of time and temperature. The time-temperature factor may be insufficient to convert even the most oil-prone organic matter to petroleum (thermal immaturity), or it may generate petroleum (thermal maturity) or generate, expel, and overheat the residual organic matter, leaving only charred carbon (thermal post maturity) in the source rock. Heating is related to burial depth, crustal tectonics, and proximity to igneous bodies.

Activation energy is the amount of energy required for a chemical reaction to proceed. In sedimentary organic matter conversion to petroleum, many reactions take place. However, because of chemical complications inherent when many different reactions occur simultaneously and sequentially, as in sedimentary organic matter conversion to petroleum, the use of experimentally derived pyrolysis activation energies is a practical way to calculate realistic conversion rates. Burnham et al. have further developed the method.

At any designated location such as a well site, the Lopatin method enhanced by the input of pyrolysis activation energies requires:

- Burial history for strata including source rock candidates.
- Measured or estimated geothermal gradient(s).
- Sediment compaction.
- Thermal conductivities for the lithostratigraphic units.

The simulated maturation process integrates temperatures through time for stratigraphic units of interest.

Thermal maturation indicators for calibrating simulation models include vitrinite reflectance (Ro), thermal alteration index (TAI) from palynomorphs, clay crystallinity, and hydrocarbon molecular ratios. Other methods for calibration include the conodont alteration index (CAI), Rock-Eval Tmax, the porphyrin maturity parameter, and biomarker ratios. Each technique has limits in sample requirements, thermal resolution, and the thermal range in which it can be applied with accuracy. Sensitivity studies demonstrate that reconstruction of the local paleoheat flow (geothermal gradients) through the geological time of interest is often the source of the greatest discrepancies between calculated maturities and thermal maturation indicators. Oils frequently carry gas chromatographic (GC) signatures useful for correlation to source rock extracts (thermal or solvent). These signatures are detectable with capillary column GC or may require the higher resolution available with mass spectrometry (GCMS). Reservoir oils and extracted bitumens contain source rock information. This "molecular stratigraphy" can reveal depositional features of petroleum source rocks, such as marine or lacustrine, clay rich or clay poor, redox potential of waters, and in some cases, the age of the source rock (using age-diagnostic biomarkers).

The petroleum yield of source rocks is related to the richness and quality of its incorporated sedimentary organic matter. Several methods have been proposed for determining potential volumetric yields and percentage of conversion of source rocks.

Bitumen generated in source rocks saturates the source rock porosity. In source rocks with sufficient quantities of generated bitumen, the primarily physical process of expulsion may follow depending upon:

- The porosity, strength, and composition (e.g., clay, carbonate, or evaporite) of the source rock and its adjoining strata.
- The concentration and distribution of kerogen (e.g., layered or disseminated).
- Whether the bitumen or mineral matrix is load bearing.
- The properties of generated products (e.g., gas pressure and viscosity).
- The heating rate.

Petroleum Source Rock Formation

Petroleum source rock formation requires both (1) sufficient biological productivity to create large quantities of organic matter and (2) suitable depositional conditions for its concentration and preservation. Creation of most hydrogen-rich, oil prone, sedimentary organic matter is ultimately attributed to photosynthesis and nutrient supply. Under appropriate conditions, phytoplankton and/or algal/bacterial mats form in lacustrine, estuarine, and marine paleoenvironments, and land plants that comprise coals form in terrestrial environments. Nutrient replenishment, sunlight, temperature, pH and Eh of waters, sediments, and soils influence biological productivity.

Source rocks form where depositional conditions shelter organic matter from destruction by oxidation and from extreme dilution by mineral input. Organic matter is preserved in oxygen-restricted sediments at depths below wave base, in stagnant waters where density or temperature-stratified water columns form, and in other areas where oxygen replenishment is low. The exposure time of hydrogen-rich organic matter to chemically destructive agents is crucial. Therefore, water column conditions may be as important as those in the upper few centimeters of sediment. Organic-rich sediments often retain millimeter-thick depositional laminae, which reflect no sediment-churning bioturbation. Bioturbation records sediment invasion mainly by animals grazing on hydrogen-ri h sedimentary organic matter.

Formation of Reservoir Rock

Reservoir rocks are sedimentary rocks formed by physical, chemical or/and biological processes operating at the surface of the earth (sea and land) and resulting in the accumulation of solid grains, mineral, lithic and bio-skeletal. They can be separated into two main groups: silico-clastics produced by hydrodynamics (rivers, waves) and carbonates related to (marine) biological production of calcium carbonate (coral reef). The pore space is either preserved from primary deposition features (inter-granular) or created by chemical dissolution (vugs, karsts) or tectonic deformation (fracture). In petroleum reservoirs, porosity is generally ranging between 50-5 % of the total rock volume. Many physico-chemical processes (diagenesis) can increase or decrease the porosity of the reservoir from the deposition time (at surface) to the depth and time of petroleum emplacement.

The porosity (pore space/bulk reservoir rock volume) controls the volume of hydrocarbon trapped in the subsurface at the time of the petroleum emplacement in the trap. For microbiological considerations, it is important to bear in mind that the dominant pore size in petroleum accumulations is in the order of 1 mm in diameter, resulting in significant capillary forces and very high specific surface area that favors the solid-fluid interactions and the formation of bio-coatings. In general porosity decreases exponentially from 50 % to 0 % (compaction cementation), between the surface and 6000 m burial, but few percent of porosity have been found at 10000 m depth. Pore types (geometry and topology) probably have a significant effect on microbial activities in the reservoir but this parameter is not fully understood.

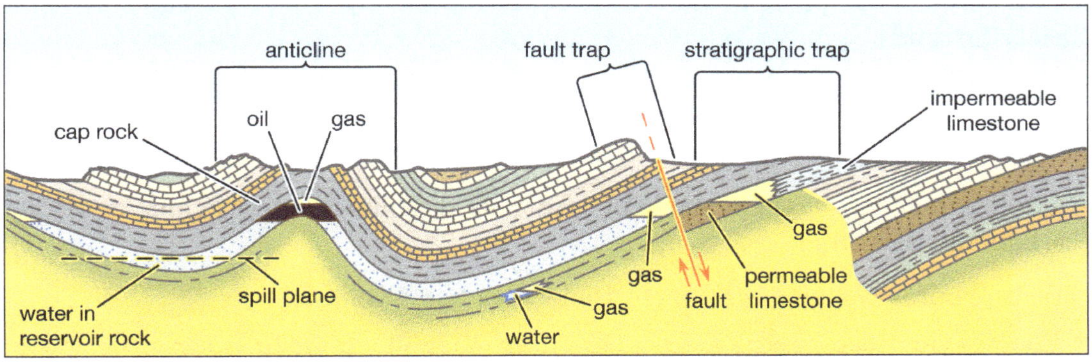

Types of Reservoir Rocks and their Properties

As a rock to be named a reservoir has to be a porous and permeable lithological structure. It encompasses sedimentary rocks. These sedimentary rocks may be made of sandstones (quartz sand or arksosic sandstone), carbonates mud or dolomite. Dolomites mostly form good reservoirs because the common reason behind it is that there is Mg, 13% smaller than Ca in a way that during dolomitization, there is a total decrease in volume of the material by 13%, here by 13% porosity is gained.

Sandstone Reservoir Rocks

The term sand refers to a specific grain with sizes between (62 μm - 2 mm). The performance of the sandstone as a reservoir rock is described by its combination of porosity and permeability depending on the degree to which the sand dominates its. The favorable texture is depicted by packaging of similar sized grains, not a combination of coarse and fine grained composition. The best sandstone reservoirs are those that are composed mainly of quartz grains of sand size of nearly equal sizes or silica cement, with minimal fragmented particles. Sandstone reservoirs are generally 25 meters thick.

Carbonate Reservoir Rocks

The most fascinating aspects of carbonate reservoir rocks are their content. Carbonates are usually made of fossils which "range from the very small single cell to the larger shelled animals". Most carbonate rocks are deposited at or in very close neighborhood to their site of creation. The "best-sorted" carbonate rocks are Oolites in which encompass grains of the same size and shapes even though Oolites are poorly sorted.

Reservoir Rock Properties, Interpretations and their Significance on a Petroleum System

Porosity of reservoir is the property that tells how porous a rock is. It is also defined as a measure of the capacity of reservoir rocks to contain or store fluids. The porosity is genetically classified basing on standard sedimentologic description of reservoir rock; there are primary and secondary porosity.

The primary porosity types are:

- Inter-particle- In this type by which rock content was quickly lost in muds and carbonate sands through compaction and cementation respectively. This type is mostly found as siliciclastic sands.
- Intra particle porosity by which the porosity is made of interiors of carbonate skeletal grains.

Secondary porosity, the porosity formed after deposition leads to other couple of reservoirs types:

- Dissolution porosity type is made of carbonate dissolution and leaching. It is also called carbonate reservoirs.
- Fracture porosity which is characterized by not being voluminous.

Porosity can also be classified basing on rock morphology. There are three types of morphologies to the pore spaces which are:

- Caternary in which the pore open to more than one throat passage.
- Cul-de-sac in which the pore open to only one throat passage.
- Closed pore in which there is no connection with other pores.

Permeability is a measure of the ability of a fluid to pass through its porous medium. Permeability is one of important to determine the effective reservoir. Porosity and permeability are two properties describing the reservoir rock capacity with regard to the fluid continence. Moreover, a reservoir rock can be porous without being permeable. For example it is said to be permeable if and only if the pores "communicate". Hence for explorationists, knowing reservoir rock permeability is a key mile stone because it is important for being used to determine if it really has sufficient commercial accumulation of oil, indeed measuring it is very difficult. The measuring of permeability can differently be understood basing on two different ways. When the porous medium is completely saturated by a single fluid, the permeability will be described absolute, become described as effective permeability when its porous medium is occupied by more than one fluid.

Other Factors Affecting the Volume of the Reservoir Rocks

- Grain size and pattern arrangement: Apart from the arrangement pattern of grains size which effect rock properties, the actual size of the grains does not affects the permeability of a neither reservoir rock nor porosity.
- Shape of the grains: Grains with high sphericity tend to pack themselves well to make a minimum pore space, the fact which increases angularity and hence pore space volume increases.

- Sorting or uniformity of size of the grains: Size of grains has an effect on reservoir properties; the more uniform the grains are sized, the great proper volume of voids spaces. Thereby mixing grains of different sizes tends to decrease total volume of void space.
- Subsequent action to the sediments (compaction): The more grains are compacted, more the volume of void spaces decreases. However the compaction of sand is less effective than the way clay does.
- How the grains were formed.

Methods Determining Rock Properties

Reservoir rock properties such as porosity and permeability are directly or indirectly measured. The direct methods consists of measuring the core sample taken from the parallel lithological area of the reservoir rock to assess them while the indirect methods consist of using data collection, well logs, seismic, production tests, etc., the porosity data are used in the basic reservoir to evaluate volumetric calculation of fluids in the reservoir and calculating fluid saturations and geologic characterization of the reservoir.

Reservoir Rock Properties Calculation Formula

Porosity: $\phi = V_p = V_B - V_S$

ϕ symbolizes porosity, V_p (volume of all pores), V_s represents volume occupied by other particles (Matrix materials), V_B represents the total volume, p represents their respective densities.

Types of Reservoirs

Siliciclastic Reservoir

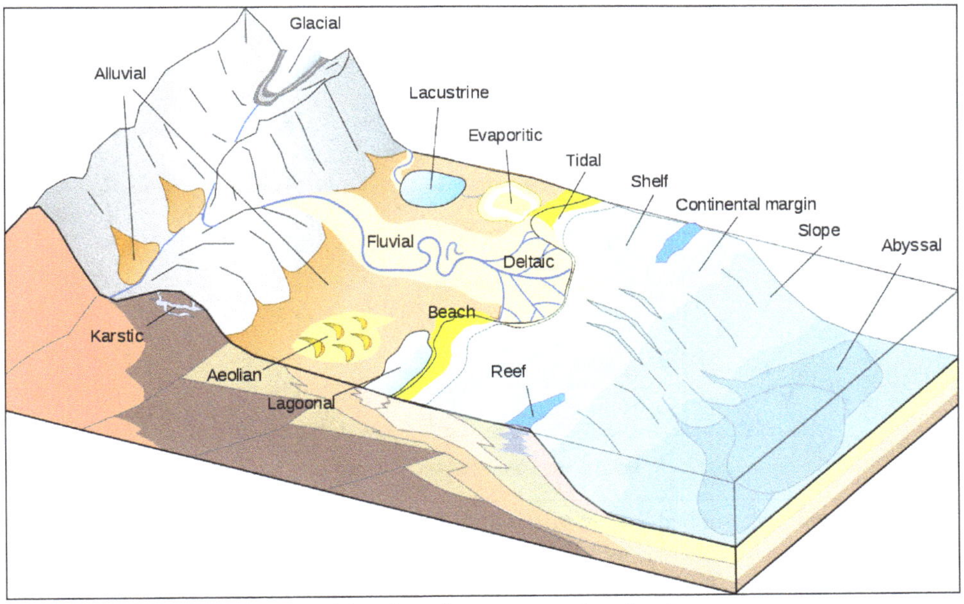

Depositional environment and grain size.

Siliciclastic sedimentary rocks are the most abundant of the sedimentary rock. They are formed from the detritus left over from the weathering of igneous, metamorphic, and older sedimentary rocks.

Shallow and Deep Marine Reservoir

Depending on the type of sedimentary rock, hydrocarbon reservoir can be divided into two groups. There are clastic sedimentary rock and non clastic sedimentary rock. On the clastic sedimentary, contained some precipitation area, as one in marine area. Rock type can be formed in deposition marine areas such as shelf sandstone and turbidity sandstone.

Shelf sandstones are formed from precipitating in the shallow marine area. Sands transported by water current from river to shallow marine area. Because of that process, has been formed sands body around the grow delta and maybe form the fan like in delta.

Turbidity sandstones: rock formed in deep marine area with rotation force existing rotation deep current so as formed the coarse layer which has interaction with shale layer in the deep marine. Shape from deposition of turbidity sandstones can be like a lens, duct, or fans.

Lacustrine Reservoir

This type of reservoir formed in basin containing water surrounded by land and initially formed by tectonic processes, volcanic, rifting, soil movement, the erosion by the wind on the coast or in land. The texture of sedimentary rocks in the environments usually granules grained and the size between 2 mm–4 mm.

Eolian Reservoir

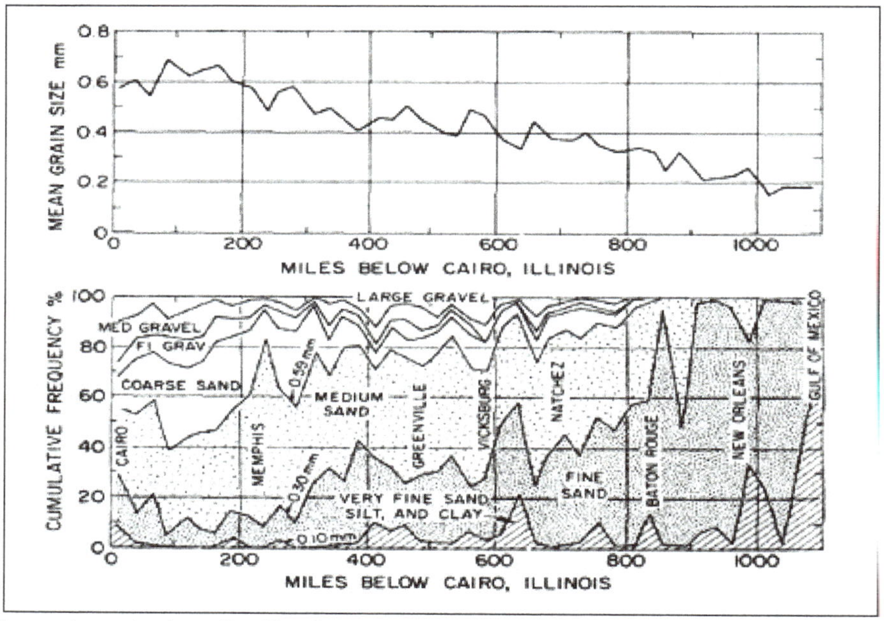

Change in grain size of bed load in Mississippi River from Cairo, Illinois, to the mouth.

Formed in large areas with the accumulation of sand deposition. The sediment resulting from wind-blown sand. The clastic texture of the environment is granules (2–4 mm) and coarse (over 2 mm).

Table: Modern dunes classification.

Name	Morphology	No. of Slip-faces	Wind Direction	Sand Supply (relative)	Relative Size
Sheet	Sheetlike with broad flat surface	None	Multidirectional	Abundant	Large, thick
Stringer	Thin, elongate strip	None	Unidirectional	Sparse	Extensive, thin
Dome	Circular or elliptical mound	None	Bidirectional	Sparse	Small, thin
Barchan	Crescent	1	Unidirectional	Sparse	Small, thin
Barchanoid ridge	Row of connected crescents	1	Unidirectional	Abundant	Large, thick
Transverse ridge	Asymmetrical ridge	1	Unidirectional	Very abundant	Large, thick
Parabolic	U-shape	1 or more	Unidirectional	Abundant	Large, thick
Linear (seif)	Symmetrical ridge	2	Bidirectional	Sparse	Extensive, thin
Reversing	Asymmetrical ridge	2	Bidirectional		Small, thin
Star	Central peak with 3 or more arms	3 or more	Multidirectional	Sparse	Small, thin
Blowout	Circular rim of depression	1 or more	Multidirectional	Variable	Small

Fluvial Reservoir

Deltaic Reservoir

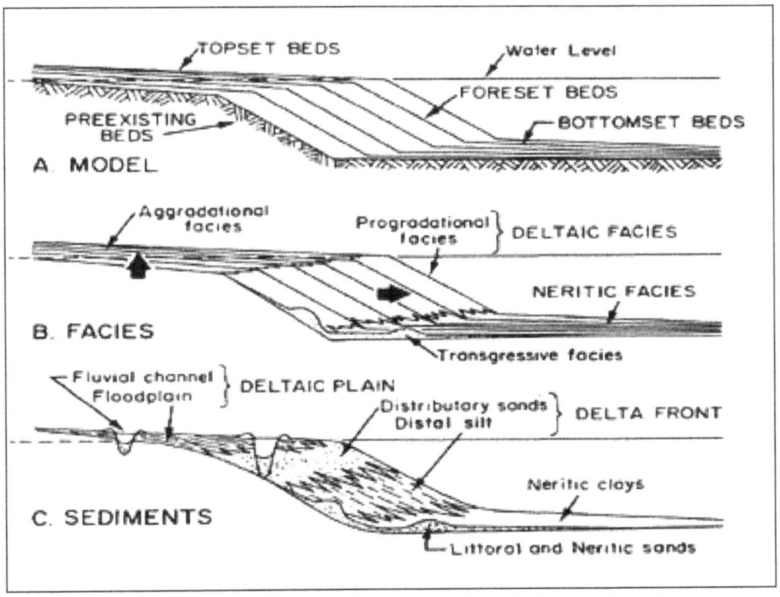

Diagrammatic cross sections of depositional units within deltas. (A) Delta Concept of Gilbert showing topset, foreset, and bottomset beds; (B) Deltaic and neritic facies from Frazier; (C) Sediment types and depositional units of an idealized delta.

On this type of reservoir, it was formed by the accumulation of lacustrine sediments. Very fine grain sediment rocks.

Carbonate Reservoir

Carbonate rocks is a sedimentary rocks with carbonate fraction more than 50%. Carbonate rocks can be used as a reservoir because of its porosity and permeability. Carbonate rocks can be classified to clastic and non clastic sediment. Its environmental formation is Tropic Ocean. The porosity concept of a carbonate rock might be a little bit more complex than other rocks, because of its vary secondary porous, from carbonate dissolution made from skeletal remain and microbe with cement. Almost all of carbonate reservoir type accumulated as a shallow marine sedimentary, except on a pelagic chalk and deep marine re-sediment reservoir carbonate in Mexico seas. Carbonate reservoir rocks can be found as a clastic limestone, carbonate framework (reef), and dolomite.

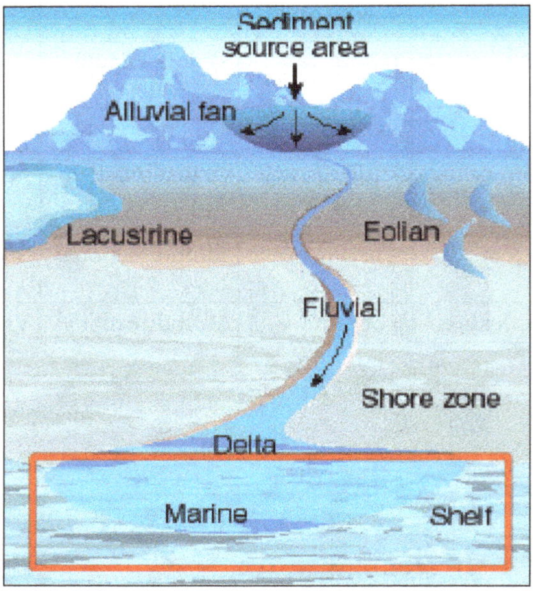

Reef Reservoir

Reef habitat in the shallow sea.

Reef is a framework made of sea organism containing skeletal, grow in shallow clean water where sunlight can reach as nutrition. Reef distribution is varies, some on the edge of the shelf and become a barrier, some scattered called patch reef. Patch reef can reach a few kilometer sizes while barrier can be elongated along the edge and limit the exposure with basin. Meanwhile the shape of a reef can be a pole (pinnacle) or lengthened (fringing). Both can become a good reservoir. Reef is a non-clastic carbonate rocks without transportation process on its formation.

Clastic Limestone

Clastic limestone usually associated with oolit and become a pretty good reservoir. Limestone associated with oolite often referred to as calcarenite. The deposition is in shallow marine environments along the coast with high energy (strong wave currents). Porosity may be extremely high because of the dissolution, but permeability is not far from 5 mD. It is called clastic because oolite associated with limestone is present through the transport process before finally being deposited.

Dolomite

Dolomite formed by processes of calcite dolomitization from other carbonates (e.g. limestone). Dolomitization occurred not long after the process of sedimentation. Dolomitization process can be reviewed as the secretion of magnesium contained in sea water into the compounds associated with carbonate. Dolomite production begins from the Pre-Cambrium, continued from Paleozoic to Mesozoic, up to Tertiary time. Calcite will be replaced with the dolomite compound that has smaller volume, so that the space between the pores of the rock grew wide. This resulted in the cavities. Calcite that has not been dolomitized will dissolve quickly and arising secondary pores, sometimes with very large size. Dolomite has its price of porosity as high as limestone porosity, which is about 20-30%. However, permeability is very high, reaching up to 5 mills Darcy to 2 Darcy. Hence, dolomite carbonate reservoirs have higher production of hydrocarbon than limestone. According Zenger estimates that 80% of carbonate reservoirs in the USA contains dolomite, while the 20% rest of it contains limestone.

Aphanitic Limestone

Aphanitic limestone formed in shallow ocean areas with low levels of energy, such as for example in the lagoon. This limestone has a very fine grain size and tight. However, the porosity and permeability was appeared by the events of dissolution and cracking that causing this type of limestone have secondary porosity. Thus, it can be a reservoir of oil.

Another Type of Reservoir

Although the porosity and permeability are poor, shale, silt stone, limestone can even act as reservoir due to fractures in the rock body (secondary porosity – secondary permeability). For example, oil field in Florence, Colorado which is having shale (Lower–Upper Cretaceous) as reservoir rock.

Then, it shows that for other than sedimentary rocks (igneous – metamorphic) could be reservoir rock if there are in fracturing state. For example, in Cuba, the oil is obtained from ultra-base igneous rock or volcanic rock that has fractured. There are eight oil fields in Cuba in 1964 that produce 710 barrels oil per day. Reservoir from this type has a very small percentage compared to the reservoir from sedimentary rock (about 1% of the overall reservoir in the world).

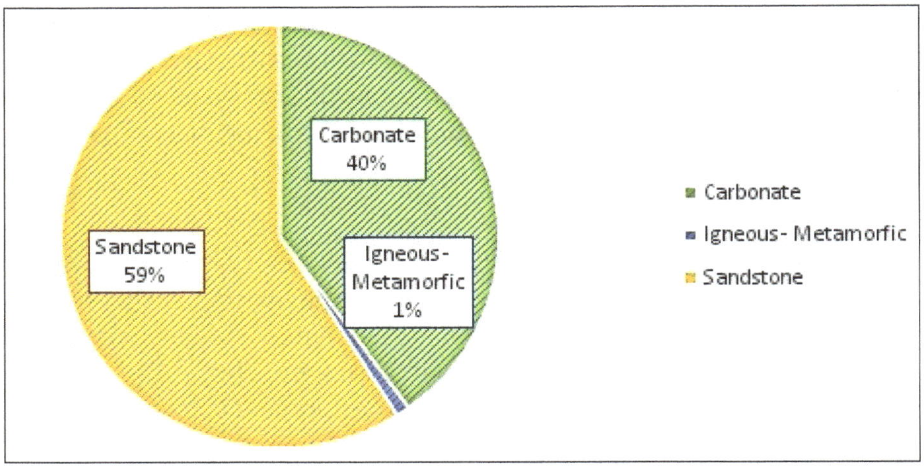

Comparison of reservoir rock types around the world.

Volcanic rocks are igneous rocks that formed on the earth surface (extrusive igneous rock). Volcanic rock can be formed from mafic minerals such as olivine, pyroxene, amphibole, and biotite, or felsic minerals such as feldspar, muscovite, and quartz. For example, oil obtained in Jatibarang (West Java, Indonesia), produced from fractures that occur in volcanic rock (tuff). Oil production from volcanic rock reservoir shows higher production in initially, and then shows a rapid decline in production.

Porosity and Permeability of Reservoir Rocks

The pores themselves may be studied by a variety of methods ranging from examination of rough or polished rock surfaces by hand lens or stereoscopic microscope, through study of thin sections using a petrological microscope or by the use of scanning electron microscope. Another effective technique of studying pore fabric is to impregnate the rock with a suitable plastic resin and then dissolve the rock itself with an appropriate solvent. Examination of the residues gives some indication, not only of the size and shape of pores, but also of the throat passages, which connect pores. The minimum size of throats and the tortuosity of pore systems are closely related to the permeability of the rock.

The classification shown in Table divides porosity types into two main varieties, which are commonly recognized. These are primary porosity fabrics, which were present during and immediately after deposition of sediment and secondary or post depositional fabrics, which formed after sedimentation by a variety of causes.

Table: Classification of Porosity Types.

	Type	Origin
Primary or depositional	(a) Intergranular or interparticle	Sedimentation
	(b) Intraparticle	

Secondary or post depositional	(c) Intercrystalline	Igneous rock, Dolomite
	(d) Fenestral	Cementation
	(e) Mouldic	Fabric-selective solution
	(f) Vuggy	Fabric-inselective solution
	(g) Fracture	Tectonic movement compaction or dehydration

Primary or Depositional Porosity

Primary or depositional porosity is that which, by definition, forms when a sediment is laid down. Two main types of primary porosity may be recognized.

Intergranular Porosity

Intergranular or interparticle porosity occurs in the spaces between the detrital grains which form the framework of sediments. This is very important porosity type in sedimentary rocks and is present initially in almost all sedimentary rocks. Intergranular porosity is generally progressively reduced by diagenesis in many carbonates, but it is dominant porosity type found in sandstones. Sediment grain size has no role on the volume of porosity in any clastic rock (sandstone or shale). A mass of spheres of uniform sorting and packing will have same porosity, regardless of size of the spheres. Porosity volume is dependent of grain packing (Cubic packing 48% porosity, Rhombohedral packing 26% porosity). Grain shape also has a role.

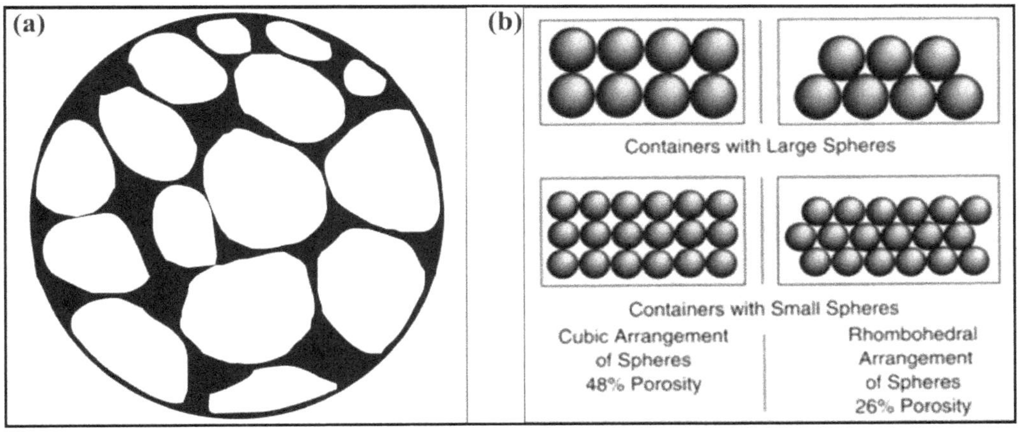

(a) Interparticle porosity; (b) Illustration showing control of grain packing on porosity irrespective of grain size.

Intergranular porosity and permeability are greatly reduced by burial diagenesis. When deposited, initial porosities may be 35% to 40% and permeabilities of several darcies. Typical oil and gas reservoirs have porosities ranging from 10% to 25% and permeabilities of a few to a few hundred millidarcies.

Intraparticle Porosity

In carbonate sands, particularly those of skeletal origin, primary porosity may be present within the detrital grains. For example, the cavities of molluscs, ammonites, corals, bryozoan and microfossils may be all classed as intraparticle primary porosity. This kind of porosity is often diminished shortly after deposition by infiltrating micrite matrix. Furthermore, the chemical instability of the carbonate host grains often leads to their intraparticle pores being modified or obliterated by subsequent diagenetic history.

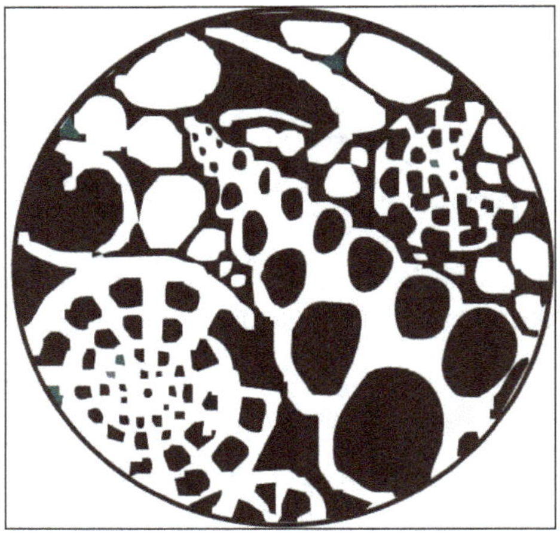

Effect of Sorting and Grain Size on Porosity and Permeability

Porosity increases with increasing sorting. In poorly sorted sandstone, finer grains occupy the interstitial spaces between the larger grains and thereby decrease porosity of sandstone. In fact, finer grains block both pores and pore throat passages. Both grain shape and roundness affect porosity. Lower the roundness of grain higher will be the porosity. This is because packing of grains become tighter with higher sphericity of grains.

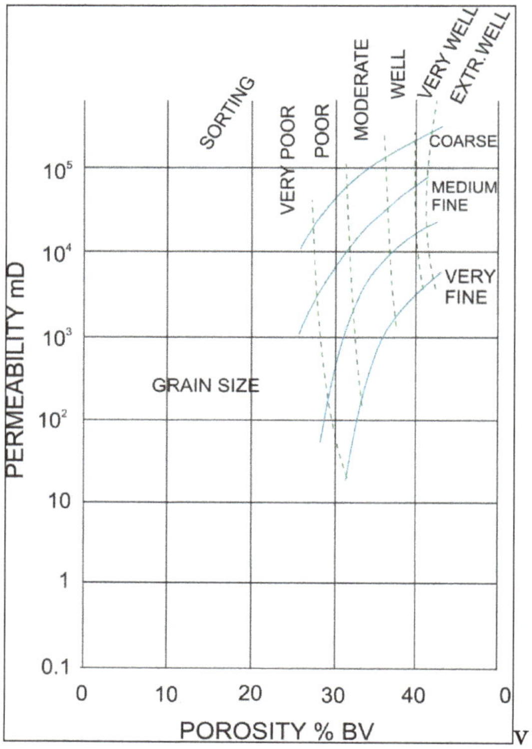

Control of grain size and grain sorting on reservoir porosity and permeability.

Table: Effect of rock properties on Porosity and Permeability.

Rock property texture	Effects on permeability and porosity
Grain size	Permeability decreases with grain size; porosity unchanged.
Sorting	Permeability and porosity decrease as sorting becomes poorer.
Packing	Although little data is available, tighter packing favors both lesser permeability and porosity.
Grain fabric	In absence of lamination, controls anisotropy of permeability, permeability is maximum parallel to mean shape fabric.
Cement	The more cement, less permeability and porosity.

Secondary or Post Depositional Porosity

Secondary porosity is that which, by definition formed after sediment was deposited. Secondary porosity is more diverse in morphology and more complex in genesis than primary porosity. The following main types of secondary porosity are recognizable.

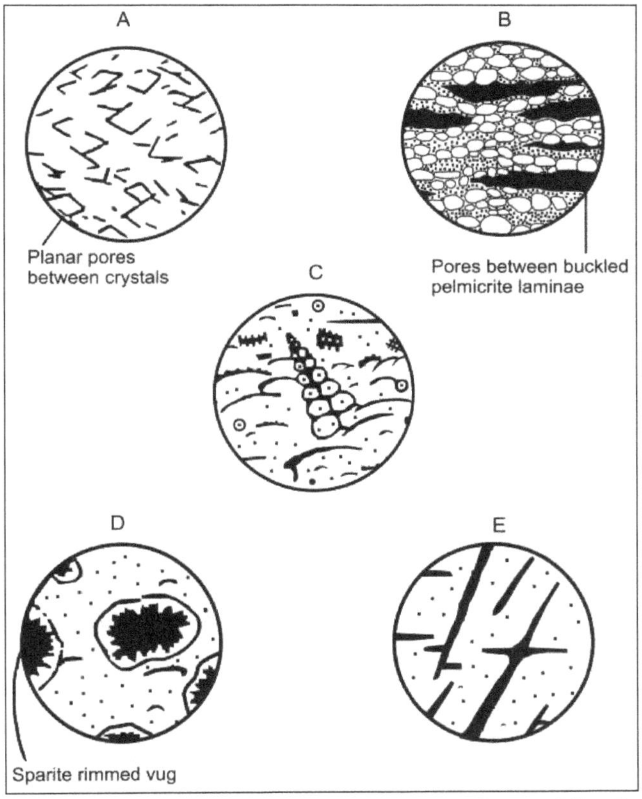

Different types of secondary porosity a. Intercrystalline, b) Fenestral, c) Moldic, d) Vuggy and e) Fracture.

Intercrystalline Porosity

Intercrystalline porosity occurs between individual crystals of a crystalline rock. It is, therefore, the typical porosity type of igneous and high-grade metamorphic rocks and of some evaporates.

Strictly speaking, such porosity is of primary origin. It is, however, most characteristics of carbonates, which have undergone crystallization, and is particularly important in recrystallized dolomites. The pores of crystalline rocks are essentially planar cavities, which intersect obliquely with one another with no constrictions of the boundaries or throats between adjacent pores.

Fenestral Porosity

The term fenestral porosity was proposed for a primary or penecontempraneous gap in rock framework, larger than grain supported interstices. This porosity type is typical of fragmental carbonate sands, where it grades into primary porosity. It is most characteristic of pellet muds and homogeneous muds of lagoonal and intertidal origin. Pene-contemporaneous dehydration, cementation, and gas generation can cause depositional laminae to buckle and generate sub horizontal fenestral pores between the laminaes.

A variety of fenestral fabric has long been known as bird's eye. This refers to isolated eyes up to a centimeter across which occurs in some lime mudstone. These structures have been attributed to organic burrows and to gas escape conduits. They are frequently infilled by crystalline calcite.

Moldic Porosity

A third type of secondary porosity, generally formed later in the history of rock than fenestral is moldic porosity. Molds are pores formed by the solution of primary depositional grains generally subsequent to some cementation. Molds are fabric selective. That is to say, solution is confined to individual particles and does not cross cut cement matrix and framework. Typically, in any one rock it is all the grains of one particular type that are dissolved. Hence, one may talk of oomoldic, pelmoldic or biomoldic porosity where there has been selective solution of ooliths, pellets or skeletal debris. The geometry and effective porosity and permeability of moldic porosity can thus be extremely varied. In an oomoldic rock, pores will be subspherical and of similar size. In biomoldic rocks, by contrast, pores may be very variable in size and shape, ranging from minute to curved planar pores where shells have dissolved, and cylinders where echinoid spines have gone into solution.

Vuggy Porosity

Vugs are a second type of pore formed by solution and like molds, they are typically found in carbonates. Vugs differ from molds though because they cross cut the primary depositional fabric of the rock. Vugs tend to be larger than molds. They are often lined by a selvedge of crystals. With increasing size, vugs grade into what is loosely termed cavernous porosity. The minimum dimension of a cavern is a pore which allows a man to enter or which, when drilled into, allows the drill string to drop by more than half a meter through the rotary table. Such cavernous porosity occurs in the Arab zone (upper Jurassic) of the abquaiq oil field in Saudi Arabia. Some of the largest cavernous oil reservoir on record occurs Texas, naturally where caverns up to five meters high were reported in the fusselman limestone of Dollarhide oil field.

Fracture Porosity

The last main type of pore to be considered is that which occurs within fractures. Fracturing, in the sense of a breaking of depositional laminations can even occur penecontemporaneous with sedimentation. This often takes the form of microfaulting caused by slumping, sliding and compaction. Fractures in plastic sediments are instantaneously sealed. In brittle rocks, however, fractures may remain open after formation, thus giving rise to fracture porosity. This porosity type characterizes rocks which are strongly lithified and is therefore, generally formed later in time than the other varieties of porosity.

It is important to note that fracture porosity is not only found in well cemented sandstones and carbonates, but may also be present in shales and igneous and metamorphic rocks. Fracture porosity can occur in a variety of ways and situations. Tectonic movements can form fracture porosity in two ways viz. tension over crests of compressional anticlines and shear failure.

Fracture porosity is also intimately associated with faulting and some oil fields show very close structural relation with individual fault system. Fracture porosity can also form from tectonic processes. It is often found immediately beneath unconformities. Here fractures, once formed by weathering, may have been enlarged by solution (especially in limestone) and get preserved without subsequent loss of porosity.

Recognition of the significance of fracture in producing fluids from high porosity low permeability formations has led to the development of artificial fracturing by explosive charges which simultaneously wedge the fractures open with sand, glass beads etc. similarly, the productivity of fractured carbonate reservoirs may be increased by the injection of acid to dissolve and enlarge the fractures.

Secondary Porosity in Sandstone

Secondary porosity is most common in carbonates. In case of sandstones, secondary porosity is common in some sandstone those have undergone long lasting burial digenesis and have lost their primary porosity. Dissolution of carbonate cements results from decarboxylation of organic matter in strata during the course of burial and organic maturation. CO_2 and carbonic acid formed thereby cause dissolution of carbonate cement that might be present in sandstone and result in formation of secondary porosity.

Capillary Pressure Curves and Pore Morphology

Capillary pressure curves provides data to determine the size of the throats connecting the pores and the percentage of total pores effective at successive levels in a reservoir. Pore throat geometry is basically the key to hydrocarbon producing potential. Fluid movement in a rock is a function of relative permeability, which in turn, for a given rock at a given time is a function of the fluid saturation in the rocks. Fluid saturations in turn are affected by pore geometry of the rock responding to reservoir pressures. Pressure is a function of height above oil – water contact, and with an increase in differential pressure, increasingly smaller pore throats are invaded by oil.

If relative permeability to oil and water is also determined from a core sample, in addition to the capillary pressure curve, a quite accurate picture can be constructed of fluid behavior in a reservoir composed of this rock. A typical relationship between relative porosity and permeability and capillary pressure, as well as the effect of the two parameters on productibility. Most reservoir rocks are naturally water wet. As oil enters within the reservoir initially it occupies the coarser pore spaces driving water out into smaller and finer pore spaces where capillary pressure is high. This is the time reservoir only yields water. Only when the replacement of water by oil reaches 20% reservoir starts yielding water plus some oil. With time reservoir turns into oil-wet from water-wet and relative permeability of reservoir with respect to water becomes zero. The reservoir now produces oil without water. The permeability relative to oil rises rapidly and the capillary pressure increases abruptly to the maximum. At this stage, oil replaces water in almost all pore spaces except fine/very fine ones where it is held by high capillary pressure. This water is referred as irreducible water and in present case, it is 30%.

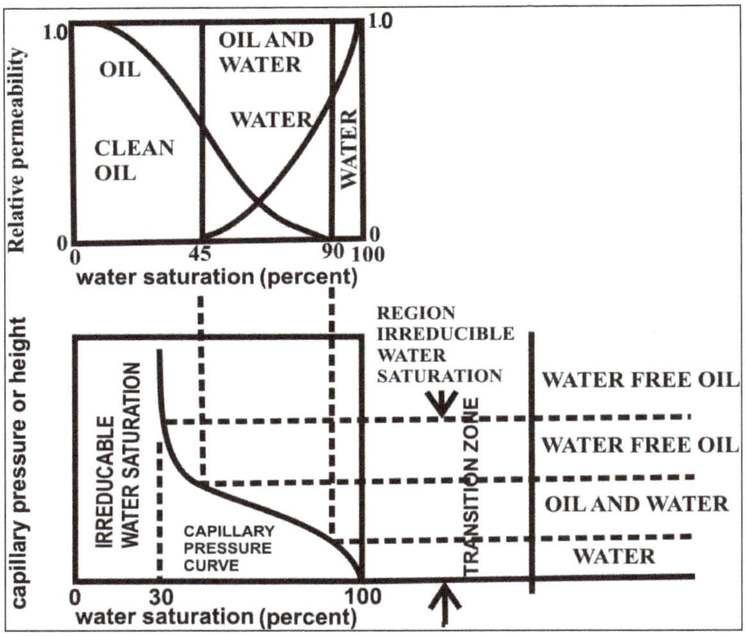

Rock productivity as determined from capillary pressure and relative permeability curves. Smallest pore throat effective at any pressure level shown on scale to right of capillary pressure curve.

The degree of sorting and skewness, which the pore system exhibits, is of more than academic interest. It should be apparent that, if a rock exhibits markedly fine skewness, excessive heights above oil water contact are required before it is capable of holding oil. Thus, even though rock may be porous, it may be incapable of holding oil. Even though rock may be porous, it may be incapable of holding oil if the closure of the potential trap is less than the required height above oil-water contact.

Permeability

Permeability measures ease of flow of a fluid through a porous media. Porosity controls volume of hydrocarbons in a reservoir, permeability controls productivity rates. The pore network is considered as a bundle of tortuous circular capillary tubes of constant diameter. Darcy's law demonstrate flow through this network,

$$Q = k\ A/\mu\ dP/dL$$

Where Q is flow in Cm^3/sec, k is permeability in darcy, A is cross-sectional area, μ is viscosity of fluid in centipoises, dP/dL is pressure gradient.

Permeability depends largely on:

- Size of pore openings.
- Degree and size of pore connectivity.
- Degree and type of cementing material between rock grains.

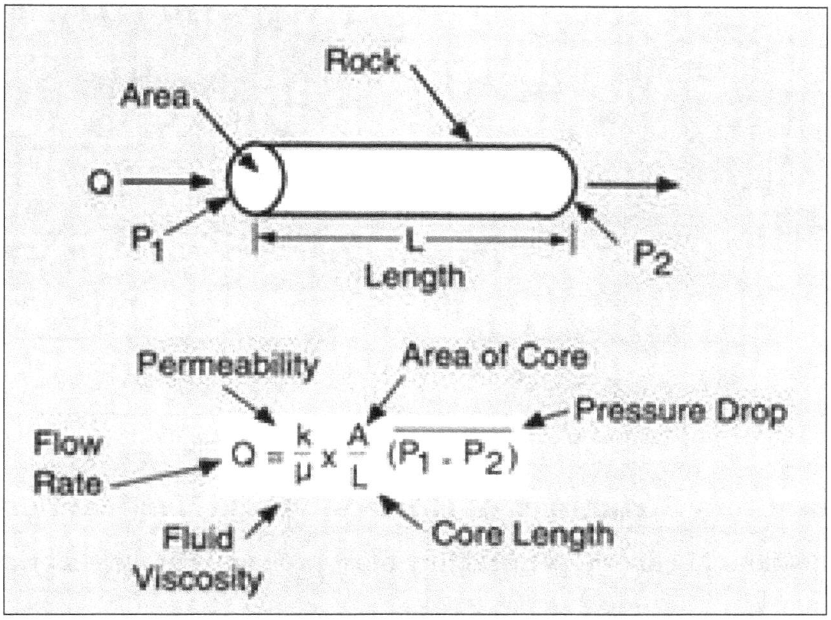

Definition of permeability from a core sample.

Absolute, Effective and Relative Permeability

Darcy's Law is valid under the following conditions:

- Entire pore spaces are filled by a single fluid i.e. saturation is 100%.
- Fluid must be a Newtonian fluid i.e. viscosity is independent of flow rate.
- Flow must be in the viscous, laminar region.

When these three conditions are met, the permeability is considered as absolute or specific permeability i.e. permeability of one phase in a single-phase system. When there is multi-fluid system, permeability of one particular fluid is its effective permeability. The effective permeability may vary from zero to 100% of absolute permeability. It depends on relative saturation of different phases, capillary pressure and surface tension and wettability of reservoir rocks.

Relative permeability is effective permeability divided by absolute permeability. Thus, relative permeability expresses the fraction of the total or absolute permeability, which a particular fluid is utilizing. Horizontal permeability is generally accepted as the rock's permeability in a more-or-less horizontal direction, while vertical permeability is generally accepted as the component perpendicular to horizontal permeability. Vertical permeability (k_v) is usually somewhat less than horizontal permeability because of the layering effect of sedimentation; i.e., clay laminae, platy minerals, etc. Horizontal permeability (k_h), measured parallel to bedding, is the major contributor of fluid flow into a typical wellbore. The ratio of k_h/k_v generally ranges from 1.5 to 3.0 but might exceed 10.0 in some reservoirs.

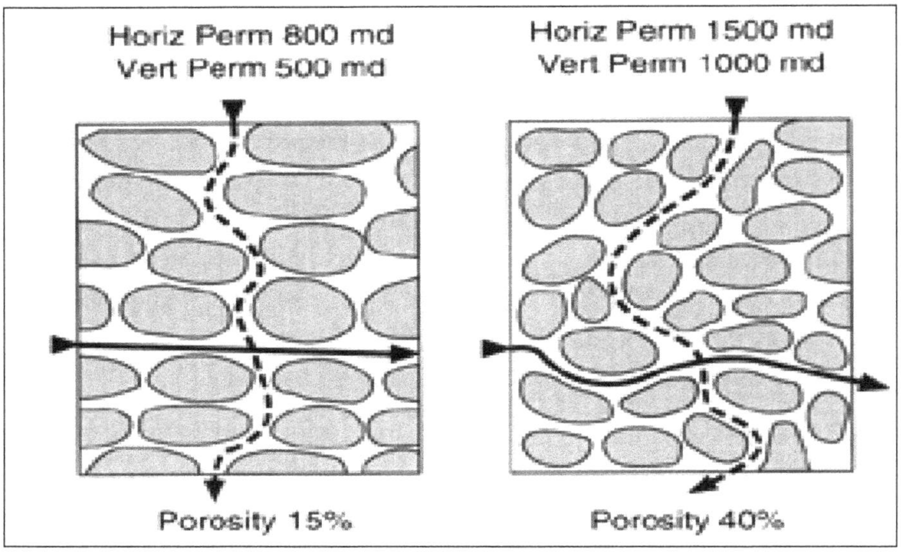

Illustration of horizontal and vertical permeability.

Effect of Sedimentary Structures of Horizontal and Vertical Permeability

- Parting lineation: Maximum permeability most probably Parallels fabric in plane of bedding.
- Cross bedding: Scant available data suggests that horizontal Permeability parallels direction of inclination and that more dip of the foreset, weaker horizontal vector of permeability.
- Ripple marks: Little data, but fine grain and more laminations combine to cause low permeability and hence ripple zones are commonly barriers to flow.
- Grooves and flutes: As judged by fabric, permeability should parallel long dimension.
- Slump structures: No data, but probably always greatly reduce horizontal permeability.
- Biogenic structures: Destroy depositional fabric and bedding and thus, drastically reduce permeability and cause minimal, if any, horizontal anisotropy of permeability. Effect on porosity is unknown but may be negligible.

Relationship between Porosity and Permeability

Even though theoretical equations may indicate a relationship between permeability and porosity, from a physical point of view such a relationship does not exist. Grain size, sorting and diagenetic products have a great effect on permeability. Figure provides a general idea on possible relationship of porosity and permeability in different types of reservoir. Whereas in an ideal clastic reservoir there may a crude relationship, between the two; in a carbonate reservoir there can be very high porosity with a low permeability and in a fractured reservoir, high permeability can be recorded with low porosity.

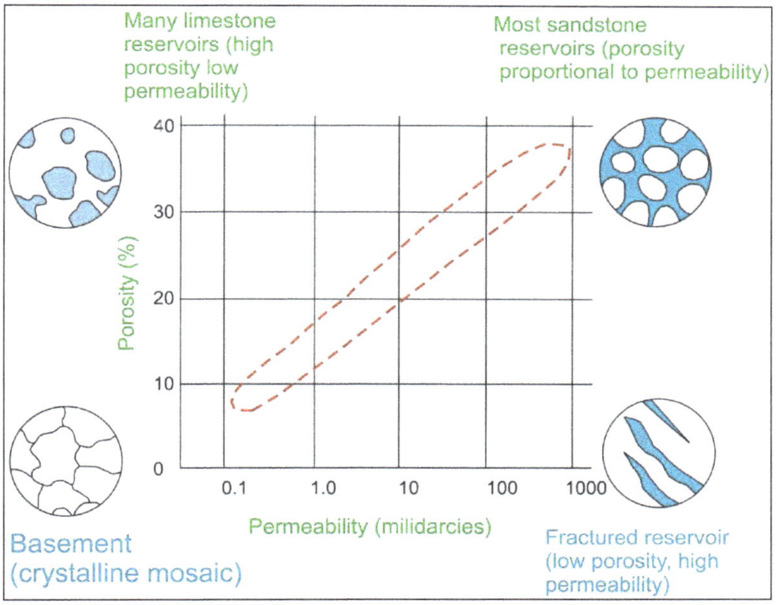

Porosity-Permeability cross-plot for different varieties of reservoirs.

Table: Comparison of porosity in sandstone and carbonate.

Aspect	Sandstone	Carbonate
Amount of primary porosity In sediment	Commonly 25-40%.	Commonly 40-70%.
Amount of ultimate porosity of rocks	Commonly half or more of initial porosity, 15-30% common.	Commonly none or only small Fraction of initial porosity, 5- 10% common in reservoir facies.
Types of primary	Almost exclusively interparticle.	Interparticle commonly pre-dominates, but intraparticle is also important.
Types of ultimate porosity	Almost exclusively primary interparticle.	Widely varied because of post depositional modifications.
Size of pores	Diameters and throat sizes closely related to sedimentary particle size and sorting.	Diameters and throat sizes commonly show little relation to sedimentary particle size or sorting.
Shapes of pores	Strong dependence on particle shape – a negative of particle.	Greatly varied, ranges from strongly dependent positive or negative of particles to form complete independent of depositional or diagenetic components.
Uniformity of size, shape and distribution	Commonly fairly uniform within homogenous body.	Variable ranging from fairly uniform to extremely heterogeneous. Even within body made up of single rock type.

Influence of diagenesis	Usually minor reduction of primary porosity by compaction and cementation.	Major influence; can create/obliterate or modify porosity.
Influence of fracturing	Generally not of major importance in reservoir properties.	Of major importance in reservoir properties if present.
Visual evaluation of porosity and permeability	Semi quantitative visual estimates commonly relative easy.	Variable, semi quantitative visual estimates range from easy to virtually impossible.
Adequacy of core analysis for reservoir evaluation	Core plugs of 1" diameter commonly adequate for matrix porosity.	Core plugs commonly inadequate, even whole cores (3"dia) may be inadequate for large pores.
Permeability and porosity interrelations	Relative consistent, commonly dependent on particle size and sorting.	Greatly varied, commonly independent of particle size and sorting.

References

- Petroleum-Reservoirs-and-Reservoir-Engineering-277352892: researchgate.net, Retrieved 05, July 2020
- Petroleum-formation, petroleum-resources: norskpetroleum.no, Retrieved 23, March 2020
- Petroleum-from-formation-to-microbiology-281330570: researchgate.net, Retrieved 20, June 2020
- Reservoir-rock, science: britannica.com, Retrieved 15, February 2020
- Reservoir-rock-properties: spe.org, Retrieved 04, May 2020

4

Hydraulic Fracturing

The method used to extract petroleum or natural gas by fracturing of rocks by a pressurized liquid is known as hydraulic fracturing. This chapter encompasses waterless fracturing, hydraulic fracturing proppants, regulation of hydraulic fracturing, hydro-slotted perforation and uses of radioactivity in oil and gas wells. The aim of this chapter is to explore the field of hydraulic fracturing.

Hydraulic Fracturing

Hydraulic fracturing (also fraccing, fracking, hydrofracturing or hydrofracking) is a well stimulation technique in which rock is fractured by a pressurized liquid. The process involves the high-pressure injection of 'fracking fluid' (primarily water, containing sand or other proppants suspended with the aid of thickening agents) into a wellbore to create cracks in the deep-rock formations through which natural gas, petroleum, and brine will flow more freely. When the hydraulic pressure is removed from the well, small grains of hydraulic fracturing proppants (either sand or aluminium oxide) hold the fractures open.

Hydraulic fracturing began as an experiment in 1947, and the first commercially successful application followed in 1950. As of 2012, 2.5 million "frac jobs" had been performed worldwide on oil and gas wells; over one million of those within the U.S. Such treatment is generally necessary to achieve adequate flow rates in shale gas, tight gas, tight oil, and coal seam gas wells. Some hydraulic fractures can form naturally in certain veins or dikes.

Hydraulic fracturing is highly controversial in many countries. Its proponents advocate the economic benefits of more extensively accessible hydrocarbons. However, opponents argue that these are outweighed by the potential environmental impacts, which include risks of ground and surface water contamination, air and noise pollution, and the triggering of earthquakes, along with the consequential hazards to public health and the environment.

Increases in seismic activity following hydraulic fracturing along dormant or previously unknown faults are sometimes caused by the deep-injection disposal of hydraulic fracturing flowback (a byproduct of hydraulically fractured wells), and produced formation brine (a byproduct of both fractured and nonfractured oil and gas wells). For these reasons, hydraulic fracturing is under international scrutiny, restricted in some countries, and banned altogether in others. Some countries have banned the practice or put moratoria in place, while others have adopted an approach involving tight regulation. The European Union is drafting regulations that would permit the controlled application of hydraulic fracturing.

Geology

Halliburton fracturing operation in the Bakken Formation, North Dakota, United States.

A fracturing operation in progress.

Mechanics

Fracturing rocks at great depth frequently becomes suppressed by pressure due to the weight of the overlying rock strata and the cementation of the formation. This suppression process is particularly significant in "tensile" (Mode 1) fractures which require the walls of the fracture to move against this pressure. Fracturing occurs when effective stress is overcome by the pressure of fluids within the rock. The minimum principal stress becomes tensile and exceeds the tensile strength of the material. Fractures formed in this way are generally oriented in a plane perpendicular to the minimum principal stress, and for this reason, hydraulic fractures in well bores can be used to determine the orientation of stresses. In natural examples, such as dikes or vein-filled fractures, the orientations can be used to infer past states of stress.

Veins

Most mineral vein systems are a result of repeated natural fracturing during periods of relatively high pore fluid pressure. The impact of high pore fluid pressure on the formation process of mineral vein systems is particularly evident in "crack-seal" veins, where the vein material is part of a series of discrete fracturing events, and extra vein material is deposited on each occasion. One example of long-term repeated natural fracturing is in the effects of seismic activity. Stress levels rise and fall episodically, and earthquakes can cause large volumes of connate water to be expelled from fluid-filled fractures. This process is referred to as "seismic pumping".

Dikes

Minor intrusions in the upper part of the crust, such as dikes, propagate in the form of fluid-filled cracks. In such cases, the fluid is magma. In sedimentary rocks with a significant water content, fluid at fracture tip will be steam.

History

Precursors

Fracturing as a method to stimulate shallow, hard rock oil wells dates back to the 1860s. Dynamite or nitroglycerin detonations were used to increase oil and natural gas production from petroleum bearing formations. On 25 April 1865, US Civil War veteran Col. Edward A. L. Roberts received a patent for an "exploding torpedo". It was employed in Pennsylvania, New York, Kentucky, and West Virginia using liquid and also, later, solidified nitroglycerin. Later still the same method was applied to water and gas wells. Stimulation of wells with acid, instead of explosive fluids, was introduced in the 1930s. Due to acid etching, fractures would not close completely resulting in further productivity increase.

Oil and Gas Wells

The relationship between well performance and treatment pressures was studied by Floyd Farris of Stanolind Oil and Gas Corporation. This study was the basis of the first hydraulic fracturing experiment, conducted in 1947 at the Hugoton gas field in Grant County of southwestern Kansas by Stanolind. For the well treatment, 1,000 US gallons (3,800 l; 830 imp gal) of gelled gasoline (essentially napalm) and sand from the Arkansas River was injected into the gas-producing limestone formation at 2,400 feet (730 m). The experiment was not very successful as deliverability of the well did not change appreciably. The process was further described by J.B. Clark of Stanolind in his paper published in 1948. A patent on this process was issued in 1949 and exclusive license was granted to the Halliburton Oil Well Cementing Company. On 17 March 1949, Halliburton performed the first two commercial hydraulic fracturing treatments in Stephens County, Oklahoma, and Archer County, Texas. Since then, hydraulic fracturing has been used to stimulate approximately one million oil and gas wells in various geologic regimes with good success.

In contrast with large-scale hydraulic fracturing used in low-permeability formations, small hydraulic fracturing treatments are commonly used in high-permeability formations to remedy "skin damage", a low-permeability zone that sometimes forms at the rock-borehole interface. In such cases the fracturing may extend only a few feet from the borehole.

In the Soviet Union, the first hydraulic proppant fracturing was carried out in 1952. Other countries in Europe and Northern Africa subsequently employed hydraulic fracturing techniques including Norway, Poland, Czechoslovakia, Yugoslavia, Hungary, Austria, France, Italy, Bulgaria, Romania, Turkey, Tunisia, and Algeria.

Massive Fracturing

Well head where fluids are injected into the ground.

Massive hydraulic fracturing (also known as high-volume hydraulic fracturing) is a technique first applied by Pan American Petroleum in Stephens County, Oklahoma, USA in 1968. The definition of massive hydraulic fracturing varies, but generally refers to treatments injecting over 150 short tons, or approximately 300,000 pounds (136 metric tonnes), of proppant.

Well head after all the hydraulic fracturing equipment has been taken off location.

American geologists gradually became aware that there were huge volumes of gas-saturated sandstones with permeability too low (generally less than 0.1 millidarcy) to recover the gas economically. Starting in 1973, massive hydraulic fracturing was used in thousands of gas wells in the San Juan Basin, Denver Basin, the Piceance Basin, and the Green River Basin, and in other hard rock formations of the western US. Other tight sandstone wells in the US made economically viable by massive hydraulic fracturing were in the Clinton-Medina Sandstone (Ohio, Pennsylvania, and New York), and Cotton Valley Sandstone (Texas and Louisiana).

Massive hydraulic fracturing quickly spread in the late 1970s to western Canada, Rotliegend and Carboniferous gas-bearing sandstones in Germany, Netherlands (onshore and offshore gas fields), and the United Kingdom in the North Sea.

Horizontal oil or gas wells were unusual until the late 1980s. Then, operators in Texas began completing thousands of oil wells by drilling horizontally in the Austin Chalk, and giving massive *slickwater* hydraulic fracturing treatments to the wellbores. Horizontal wells proved much more effective than vertical wells in producing oil from tight chalk; sedimentary beds are usually nearly horizontal, so horizontal wells have much larger contact areas with the target formation.

Shales

Hydraulic fracturing of shales goes back at least to 1965, when some operators in the Big Sandy gas field of eastern Kentucky and southern West Virginia started hydraulically fracturing the Ohio Shale and Cleveland Shale, using relatively small fracs. The frac jobs generally increased production, especially from lower-yielding wells.

In 1976, the United States government started the Eastern Gas Shales Project, which included numerous public-private hydraulic fracturing demonstration projects. During the same period, the Gas Research Institute, a gas industry research consortium, received approval for research and funding from the Federal Energy Regulatory Commission.

In 1997, Nick Steinsberger, an engineer of Mitchell Energy (now part of Devon Energy), applied the slickwater fracturing technique, using more water and higher pump pressure than previous fracturing techniques, which was used in East Texas by Union Pacific Resources (now part of Anadarko Petroleum Corporation), in the Barnett Shale of north Texas. In 1998, the new technique proved to be successful when the first 90 days gas production from the well called S.H. Griffin No. 3 exceeded production of any of the company's previous wells. This new completion technique made gas extraction widely economical in the Barnett Shale, and was later applied to other shales. George P. Mitchell has been called the "father of fracking" because of his role in applying it in shales. The first horizontal well in the Barnett Shale was drilled in 1991, but was not widely done in the Barnett until it was demonstrated that gas could be economically extracted from vertical wells in the Barnett.

As of 2013, massive hydraulic fracturing is being applied on a commercial scale to shales in the United States, Canada, and China. Several additional countries are planning to use hydraulic fracturing.

Process

According to the United States Environmental Protection Agency (EPA), hydraulic fracturing is a process to stimulate a natural gas, oil, or geothermal well to maximize extraction. The EPA defines the broader process to include acquisition of source water, well construction, well stimulation, and waste disposal.

Method

A hydraulic fracture is formed by pumping fracturing fluid into a wellbore at a rate sufficient to increase pressure at the target depth (determined by the location of the well casing perforations), to exceed that of the fracture *gradient* (pressure gradient) of the rock. The fracture gradient is defined as pressure increase per unit of depth relative to density, and is usually measured in pounds per square inch, per square foot, or bars. The rock cracks, and the fracture fluid permeates the rock extending the crack further, and further, and so on. Fractures are localized as pressure drops off with the rate of frictional loss, which is relevant to the distance from the well. Operators typically try to maintain "fracture width", or slow its decline following treatment, by introducing a proppant into the injected fluid – a material such as grains of sand, ceramic, or other particulate, thus preventing the fractures from closing when injection is stopped and pressure removed. Consideration of proppant strength and prevention of proppant failure becomes more important at greater depths where pressure and stresses on fractures are higher. The propped fracture is permeable enough to allow the flow of gas, oil, salt water and hydraulic fracturing fluids to the well.

During the process, fracturing fluid leakoff (loss of fracturing fluid from the fracture channel into the surrounding permeable rock) occurs. If not controlled, it can exceed 70% of the injected volume. This may result in formation matrix damage, adverse formation fluid interaction, and altered fracture geometry, thereby decreasing efficiency.

The location of one or more fractures along the length of the borehole is strictly controlled by various methods that create or seal holes in the side of the wellbore. Hydraulic fracturing is performed in cased wellbores, and the zones to be fractured are accessed by perforating the casing at those locations.

Hydraulic-fracturing equipment used in oil and natural gas fields usually consists of a slurry blender, one or more high-pressure, high-volume fracturing pumps (typically powerful triplex or quintuplex pumps) and a monitoring unit. Associated equipment includes fracturing tanks, one or more units for storage and handling of proppant, high-pressure treating iron, a chemical additive unit (used to accurately monitor chemical addition), low-pressure flexible hoses, and many gauges and meters for flow rate, fluid density, and treating pressure. Chemical additives are typically 0.5% percent of the total fluid volume. Fracturing equipment operates over a range of pressures and injection rates, and can reach up to 100 megapascals (15,000 psi) and 265 litres per second (9.4 cu ft/s) (100 barrels per minute).

Well Types

A distinction can be made between conventional, low-volume hydraulic fracturing, used to stimulate high-permeability reservoirs for a single well, and unconventional, high-volume hydraulic fracturing, used in the completion of tight gas and shale gas wells. High-volume hydraulic fracturing usually requires higher pressures than low-volume fracturing; the higher pressures are needed to push out larger volumes of fluid and proppant that extend farther from the borehole.

Horizontal drilling involves wellbores with a terminal drillhole completed as a "lateral" that extends parallel with the rock layer containing the substance to be extracted. For example, laterals extend 1,500 to 5,000 feet (460 to 1,520 m) in the Barnett Shale basin in Texas, and up to 10,000 feet (3,000 m) in the Bakken formation in North Dakota. In contrast, a vertical well only accesses the thickness of the rock layer, typically 50–300 feet (15–91 m). Horizontal drilling reduces surface disruptions as fewer wells are required to access the same volume of rock.

Drilling often plugs up the pore spaces at the wellbore wall, reducing permeability at and near the wellbore. This reduces flow into the borehole from the surrounding rock formation, and partially seals off the borehole from the surrounding rock. Low-volume hydraulic fracturing can be used to restore permeability.

Fracturing Fluids

Water tanks preparing for hydraulic fracturing.

The main purposes of fracturing fluid are to extend fractures, add lubrication, change gel strength, and to carry proppant into the formation. There are two methods of transporting proppant in the fluid – high-rate and high-viscosity. High-viscosity fracturing tends to cause large dominant fractures, while high-rate (slickwater) fracturing causes small spread-out micro-fractures.

Water-soluble gelling agents (such as guar gum) increase viscosity and efficiently deliver proppant into the formation.

Process of mixing water with hydraulic fracturing fluids to be injected into the ground.

Fluid is typically a slurry of water, proppant, and chemical additives. Additionally, gels, foams, and compressed gases, including nitrogen, carbon dioxide and air can be injected. Typically, 90% of the fluid is water and 9.5% is sand with chemical additives accounting to about 0.5%. However, fracturing fluids have been developed using liquefied petroleum gas (LPG) and propane in which water is unnecessary.

The proppant is a granular material that prevents the created fractures from closing after the fracturing treatment. Types of proppant include silica sand, resin-coated sand, bauxite, and man-made ceramics. The choice of proppant depends on the type of permeability or grain strength needed. In some formations, where the pressure is great enough to crush grains of natural silica sand, higher-strength proppants such as bauxite or ceramics may be used. The most commonly used proppant is silica sand, though proppants of uniform size and shape, such as a ceramic proppant, are believed to be more effective.

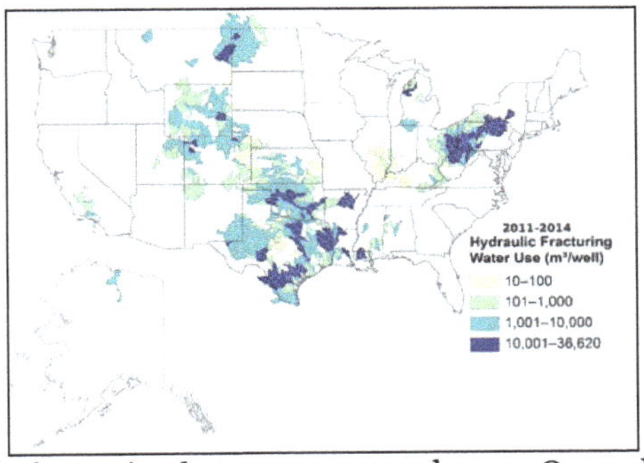

USGS map of water use from hydraulic fracturing between 2011 and 2014. One cubic meter of water is 264.172 gallons.

The fracturing fluid varies depending on fracturing type desired, and the conditions of specific wells being fractured, and water characteristics. The fluid can be gel, foam, or slickwater-based. Fluid choices are tradeoffs: more viscous fluids, such as gels, are better at keeping proppant in suspension; while less-viscous and lower-friction fluids, such as slickwater, allow fluid to be pumped at higher rates, to create fractures farther out from the wellbore. Important material properties of the fluid include viscosity, pH, various rheological factors, and others.

Water is mixed with sand and chemicals to create fracking fluid. Approximately 40,000 gallons of chemicals are used per fracturing. A typical fracture treatment uses between 3 and 12 additive chemicals. Although there may be unconventional fracturing fluids, typical chemical additives can include one or more of the following:

- Acids—hydrochloric acid or acetic acid is used in the pre-fracturing stage for cleaning the perforations and initiating fissure in the near-wellbore rock.
- Sodium chloride (salt)—delays breakdown of gel polymer chains.
- Polyacrylamide and other friction reducers decrease turbulence in fluid flow and pipe friction, thus allowing the pumps to pump at a higher rate without having greater pressure on the surface.
- Ethylene glycol—prevents formation of scale deposits in the pipe.
- Borate salts—used for maintaining fluid viscosity during the temperature increase.
- Sodium and potassium carbonates—used for maintaining effectiveness of crosslinkers.
- Anaerobic, Biocide, BIO—Glutaraldehyde used as disinfectant of the water (bacteria elimination).
- Guar gum and other water-soluble gelling agents—increases viscosity of the fracturing fluid to deliver proppant into the formation more efficiently.
- Citric acid—used for corrosion prevention.
- Isopropanol—used to winterize the chemicals to ensure it doesn't freeze.

The most common chemical used for hydraulic fracturing in the United States in 2005–2009 was methanol, while some other most widely used chemicals were isopropyl alcohol, 2-butoxyethanol, and ethylene glycol.

Typical fluid types are:

- Conventional linear gels. These gels are cellulose derivative (carboxymethyl cellulose, hydroxyethyl cellulose, carboxymethyl hydroxyethyl cellulose, hydroxypropyl cellulose, hydroxyethyl methyl cellulose), guar or its derivatives (hydroxypropyl guar, carboxymethyl hydroxypropyl guar), mixed with other chemicals.
- Borate-crosslinked fluids. These are guar-based fluids cross-linked with boron ions (from aqueous borax/boric acid solution). These gels have higher viscosity at pH 9 onwards and are used to carry proppant. After the fracturing job, the pH is reduced to 3–4 so that the cross-links are broken, and the gel is less viscous and can be pumped out.

- Organometallic-crosslinked fluids - zirconium, chromium, antimony, titanium salts - are known to crosslink guar-based gels. The crosslinking mechanism is not reversible, so once the proppant is pumped down along with cross-linked gel, the fracturing part is done. The gels are broken down with appropriate breakers.
- Aluminium phosphate-ester oil gels. Aluminium phosphate and ester oils are slurried to form cross-linked gel. These are one of the first known gelling systems.

For slickwater fluids the use of sweeps is common. Sweeps are temporary reductions in the proppant concentration, which help ensure that the well is not overwhelmed with proppant. As the fracturing process proceeds, viscosity-reducing agents such as oxidizers and enzyme breakers are sometimes added to the fracturing fluid to deactivate the gelling agents and encourage flowback. Such oxidizer react with and break down the gel, reducing the fluid's viscosity and ensuring that no proppant is pulled from the formation. An enzyme acts as a catalyst for breaking down the gel. Sometimes pH modifiers are used to break down the crosslink at the end of a hydraulic fracturing job, since many require a pH buffer system to stay viscous. At the end of the job, the well is commonly flushed with water under pressure (sometimes blended with a friction reducing chemical.) Some (but not all) injected fluid is recovered. This fluid is managed by several methods, including underground injection control, treatment, discharge, recycling, and temporary storage in pits or containers. New technology is continually developing to better handle waste water and improve re-usability.

Fracture Monitoring

Measurements of the pressure and rate during the growth of a hydraulic fracture, with knowledge of fluid properties and proppant being injected into the well, provides the most common and simplest method of monitoring a hydraulic fracture treatment. This data along with knowledge of the underground geology can be used to model information such as length, width and conductivity of a propped fracture.

Injection of radioactive tracers along with the fracturing fluid is sometimes used to determine the injection profile and location of created fractures.Radiotracers are selected to have the readily detectable radiation, appropriate chemical properties, and a half life and toxicity level that will minimize initial and residual contamination. Radioactive isotopes chemically bonded to glass (sand) and/or resin beads may also be injected to track fractures. For example, plastic pellets coated with 10 GBq of Ag-110mm may be added to the proppant, or sand may be labelled with Ir-192, so that the proppant's progress can be monitored. Radiotracers such as Tc-99m and I-131 are also used to measure flow rates. The Nuclear Regulatory Commission publishes guidelines which list a wide range of radioactive materials in solid, liquid and gaseous forms that may be used as tracers and limit the amount that may be used per injection and per well of each radionuclide.

A new technique in well-monitoring involves fiber-optic cables outside the casing. Using the fiber optics, temperatures can be measured every foot along the well - even while the wells are being fracked and pumped. By monitoring the temperature of the well, engineers can determine how much fracking fluid different parts of the well use as well as how much natural gas they collect.

Microseismic Monitoring

For more advanced applications, microseismic monitoring is sometimes used to estimate the size and orientation of induced fractures. Microseismic activity is measured by placing an array of geophones in a nearby wellbore. By mapping the location of any small seismic events associated with the growing fracture, the approximate geometry of the fracture is inferred. Tiltmeter arrays deployed on the surface or down a well provide another technology for monitoring strain.

Microseismic mapping is very similar geophysically to seismology. In earthquake seismology, seismometers scattered on or near the surface of the earth record S-waves and P-waves that are released during an earthquake event. This allows for motion along the fault plane to be estimated and its location in the earth's subsurface mapped. Hydraulic fracturing, an increase in formation stress proportional to the net fracturing pressure, as well as an increase in pore pressure due to leakoff. Tensile stresses are generated ahead of the fracture's tip, generating large amounts of shear stress. The increases in pore water pressure and in formation stress combine and affect weaknesses near the hydraulic fracture, like natural fractures, joints, and bedding planes.

Different methods have different location errors and advantages. Accuracy of microseismic event mapping is dependent on the signal-to-noise ratio and the distribution of sensors. Accuracy of events located by seismic inversion is improved by sensors placed in multiple azimuths from the monitored borehole. In a downhole array location, accuracy of events is improved by being close to the monitored borehole (high signal-to-noise ratio).

Monitoring of microseismic events induced by reservoir stimulation has become a key aspect in evaluation of hydraulic fractures, and their optimization. The main goal of hydraulic fracture monitoring is to completely characterize the induced fracture structure, and distribution of conductivity within a formation. Geomechanical analysis, such as understanding a formations material properties, in-situ conditions, and geometries, helps monitoring by providing a better definition of the environment in which the fracture network propagates. The next task is to know the location of proppant within the fracture and the distribution of fracture conductivity. This can be monitored using multiple types of techniques to finally develop a reservoir model than accurately predicts well performance.

Horizontal Completions

Since the early 2000s, advances in drilling and completion technology have made horizontal wellbores much more economical. Horizontal wellbores allow far greater exposure to a formation than conventional vertical wellbores. This is particularly useful in shale formations which do not have sufficient permeability to produce economically with a vertical well. Such wells, when drilled onshore, are now usually hydraulically fractured in a number of stages, especially in North America. The type of wellbore completion is used to determine how many times a formation is fractured, and at what locations along the horizontal section.

In North America, shale reservoirs such as the Bakken, Barnett, Montney, Haynesville, Marcellus, and most recently the Eagle Ford, Niobrara and Utica shales are drilled horizontally through the producing interval(s), completed and fractured. The method by which the fractures are placed along the wellbore is most commonly achieved by one of two methods, known as "plug and perf" and "sliding sleeve".

The wellbore for a plug-and-perf job is generally composed of standard steel casing, cemented or uncemented, set in the drilled hole. Once the drilling rig has been removed, a wireline truck is used to perforate near the bottom of the well, and then fracturing fluid is pumped. Then the wireline truck sets a plug in the well to temporarily seal off that section so the next section of the wellbore can be treated. Another stage is pumped, and the process is repeated along the horizontal length of the wellbore.

The wellbore for the sliding sleeve technique is different in that the sliding sleeves are included at set spacings in the steel casing at the time it is set in place. The sliding sleeves are usually all closed at this time. When the well is due to be fractured, the bottom sliding sleeve is opened using one of several activation techniques and the first stage gets pumped. Once finished, the next sleeve is opened, concurrently isolating the previous stage, and the process repeats. For the sliding sleeve method, wireline is usually not required.

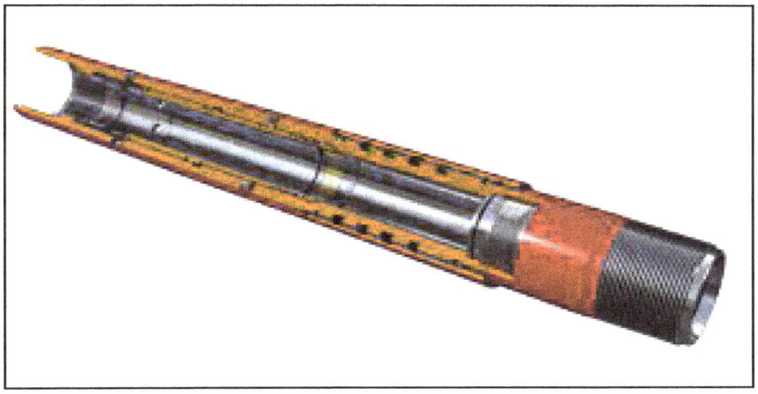

Sleeves.

These completion techniques may allow for more than 30 stages to be pumped into the horizontal section of a single well if required, which is far more than would typically be pumped into a vertical well that had far fewer feet of producing zone exposed.

Uses

Hydraulic fracturing is used to increase the rate at which fluids, such as petroleum, water, or natural gas can be recovered from subterranean natural reservoirs. Reservoirs are typically porous sandstones, limestones or dolomite rocks, but also include "unconventional reservoirs" such as shale rock or coal beds. Hydraulic fracturing enables the extraction of natural gas and oil from rock formations deep below the earth's surface (generally 2,000–6,000 m (5,000–20,000 ft)), which is greatly below typical groundwater reservoir levels. At such depth, there may be insufficient permeability or reservoir pressure to allow natural gas and oil to flow from the rock into the wellbore at high economic return. Thus, creating conductive fractures in the rock is instrumental in extraction from naturally impermeable shale reservoirs. Permeability is measured in the microdarcy to nanodarcy range. Fractures are a conductive path connecting a larger volume of reservoir to the well. So-called "super fracking," creates cracks deeper in the rock formation to release more oil and gas, and increases efficiency. The yield for typical shale bores generally falls off after the first year or two, but the peak producing life of a well can be extended to several decades.

While the main industrial use of hydraulic fracturing is in stimulating production from oil and gas wells, hydraulic fracturing is also applied:

- To stimulate groundwater wells.

- To precondition or induce rock cave-ins mining.
- As a means of enhancing waste remediation, usually hydrocarbon waste or spills.
- To dispose waste by injection deep into rock.
- To measure stress in the Earth.
- For electricity generation in enhanced geothermal systems.
- To increase injection rates for geologic sequestration of CO_2.

Since the late 1970s, hydraulic fracturing has been used, in some cases, to increase the yield of drinking water from wells in a number of countries, including the United States, Australia, and South Africa.

Economic Effects

Hydraulic fracturing has been seen as one of the key methods of extracting unconventional oil and unconventional gas resources. According to the International Energy Agency, the remaining technically recoverable resources of shale gas are estimated to amount to 208 trillion cubic metres (7,300 trillion cubic feet), tight gas to 76 trillion cubic metres (2,700 trillion cubic feet), and coalbed methane to 47 trillion cubic metres (1,700 trillion cubic feet). As a rule, formations of these resources have lower permeability than conventional gas formations. Therefore, depending on the geological characteristics of the formation, specific technologies such as hydraulic fracturing are required. Although there are also other methods to extract these resources, such as conventional drilling or horizontal drilling, hydraulic fracturing is one of the key methods making their extraction economically viable. The multi-stage fracturing technique has facilitated the development of shale gas and light tight oil production in the United States and is believed to do so in the other countries with unconventional hydrocarbon resources.

Some studies call into question the claim that hydraulic fracturing of shale gas wells has a significant macro-economic impact. A study released in the beginning of 2014 by the Institute for Sustainable Development and International Relations states that, on the long-term as well as on the short-run, the "shale gas revolution" due to hydraulic fracturing in the United States has had very little impact on economic growth and competitiveness. The same report concludes that in Europe, using hydraulic fracturing would have very little advantage in terms of competitiveness and energy security. Indeed, for the period 2030-2035, shale gas is estimated to cover 3 to 10% of EU projected energy demand, which is not enough to have a significant impact on energetic independence and competitiveness.

Research suggests that hydraulic fracturing wells have an adverse impact on agricultural productivity in the vicinity of the wells. One paper found "that productivity of an irrigated crop decreases by 5% when a well is drilled during the agriculturally active months within 11-20 km radius of a producing township. This effect becomes smaller and weaker as the distance between township and wells increases." The findings imply that the introduction of fracking wells to Alberta cost the province $3.3 million in 2014 due to the decline in the crop productivity.

Public Debate

Poster against hydraulic fracturing in Vitoria-Gasteiz, Spain, October 2012.

Politics and Public Policy

An anti-fracking movement has emerged both internationally with involvement of international environmental organizations and nations such as France and locally in affected areas such as Balcombe in Sussex where the Balcombe drilling protest was in progress during mid-2013. The considerable opposition against hydraulic fracturing activities in local townships in the United States has led companies to adopt a variety of public relations measures to reassure the public, including the employment of former military personnel with training in psychological warfare operations. According to Matt Pitzarella, the communications director at Range Resources, employees trained in the Middle East have been valuable to Range Resources in Pennsylvania, when dealing with emotionally charged township meetings and advising townships on zoning and local ordinances dealing with hydraulic fracturing.

There have been many protests directed at hydraulic fracturing. For example, ten people were arrested in 2013 during an anti-fracking protest near New Matamoras, Ohio, after they illegally entered a development zone and latched themselves to drilling equipment. Though usually non-violent, some protestors use violence and intimidation. In northwest Pennsylvania, there was a drive-by shooting at a well site, in which someone shot two rounds of a small-caliber rifle in the direction of a drilling rig, before shouting profanities at the site and fleeing the scene. In Washington County, Pennsylvania, a contractor working on a gas pipeline found a pipe bomb that had been placed where a pipeline was to be constructed, which local authorities said would have caused a "catastrophe" had they not discovered and detonated it.

In 2014 a number of European officials suggested that several major European protests against fracking (with mixed success in Lithuania and Ukraine) may be partially sponsored by Gazprom, Russia's state-controlled gas company. The New York Times suggested that Russia saw its natural gas exports to Europe as a key element of its geopolitical influence, and that this market would diminish if fracking is adopted in Eastern Europe, as it opens up significant shale gas reserves in the region. Russian officials have on numerous occasions made public statements to the effect that fracking "poses a huge environmental problem".

Documentary Films

Josh Fox's 2010 Academy Award nominated film Gasland became a center of opposition to hydraulic fracturing of shale. The movie presented problems with groundwater contamination near well sites in Pennsylvania, Wyoming, and Colorado.*Energy in Depth*, an oil and gas industry lobbying group, called the film's facts into question. In response, a rebuttal of *Energy in Depth's* claims of inaccuracy was posted on *Gasland's* website.

The Director of the Colorado Oil and Gas Conservation Commission (COGCC) offered to be interviewed as part of the film if he could review what was included from the interview in the final film but Fox declined the offer.Exxon Mobil, Chevron Corporation and ConocoPhillips aired advertisements during 2011 and 2012 that claimed to describe the economic and environmental benefits of natural gas and argue that hydraulic fracturing was safe.

The film Promised Land, starring Matt Damon, takes on hydraulic fracturing. The gas industry is making plans to try to counter the film's criticisms of hydraulic fracturing with informational flyers, and Twitter and Facebook posts.

In January 2013 Northern Irish journalist and filmmaker Phelim McAleer released a crowdfunded documentary called FrackNation as a response to the statements made by Fox in *Gasland*. *FrackNation* premiered on Mark Cuban's AXS TV. The premiere corresponded with the release of *Promised Land*.

In April 2013, Josh Fox released *Gasland 2*, a documentary that states that the gas industry's portrayal of natural gas as a clean and safe alternative to oil is a myth, and that hydraulically fractured wells inevitably leak over time, contaminating water and air, hurting families, and endangering the earth's climate with the potent greenhouse gas methane.

In 2014, Vido Innovations released the documentary *The Ethics of Fracking*. The film covers the politics, spiritual, scientific, medical and professional points of view on hydraulic fracturing. It also digs into the way the gas industry portrays fracking in their advertising.

In 2015, the Canadian documentary film Fractured Land had its world premiere at the Hot Docs Canadian International Documentary Festival.

Research Issues

Typically the funding source of the research studies is a focal point of controversy. Concerns have been raised about research funded by foundations and corporations, or by environmental groups, which can at times lead to at least the appearance of unreliable studies. Several organizations, researchers, and media outlets have reported difficulty in conducting and reporting the results of studies on hydraulic fracturing due to industry and governmental pressure, and expressed concern over possible censoring of environmental reports. Some have argued there is a need for more research into the environmental and health effects of the technique.

However, it is important to note that many of the most-cited studies over the last decade are either government estimates, environmentalist group reports, or peer-reviewed papers from academic scientists, including a 2013 EPA report that significantly lowered estimates of methane leakage compared to previous estimates, and a study commissioned by the environmentalist group Environmental Defense Fund and published in the Proceedings of the National Academy of Sciences, which similarly showed that the environmental effects of natural gas production were overestimated. Similarly, a study from the environmentalist Natural Resources Defense Council was cited to show that a previous highly cited study "significantly overestimate[s] the fugitive emissions associated with unconventional gas extraction."

Health Risks

There is concern over the possible adverse public health implications of hydraulic fracturing activity. A 2013 review on shale gas production in the United States stated, "with increasing numbers of drilling sites, more people are at risk from accidents and exposure to harmful substances used at fractured wells." A 2011 hazard assessment recommended full disclosure of chemicals used for hydraulic fracturing and drilling as many have immediate health effects, and many may have long-term health effects.

In June 2014 Public Health England published a review of the potential public health impacts of exposures to chemical and radioactive pollutants as a result of shale gas extraction in the UK, based on the examination of literature and data from countries where hydraulic fracturing already occurs. The executive summary of the report stated: "An assessment of the currently available evidence indicates that the potential risks to public health from exposure to the emissions associated with shale gas extraction will be low if the operations are properly run and regulated. Most evidence suggests that contamination of groundwater, if it occurs, is most likely to be caused by leakage through the vertical borehole. Contamination of groundwater from the underground hydraulic fracturing process itself (ie the fracturing of the shale) is unlikely. However, surface spills of hydraulic fracturing fluids or wastewater may affect groundwater, and emissions to air also have the potential to impact on health. Where potential risks have been identified in the literature, the reported problems are typically a result of operational failure and a poor regulatory environment."

A 2012 report prepared for the European Union Directorate-General for the Environment identified potential risks to humans from air pollution and ground water contamination posed by hydraulic fracturing. This led to a series of recommendations in 2014 to mitigate these concerns. A 2012 guidance for pediatric nurses in the US said that hydraulic fracturing had a potential negative impact on public health and that pediatric nurses should be prepared to gather information on such topics so as to advocate for improved community health.

Environmental Impacts

The potential environmental impacts of hydraulic fracturing include air emissions and climate change, high water consumption, water contamination, land use, risk of earthquakes, noise pollution, and health effects on humans. Air emissions are primarily methane that escapes from wells, along with industrial emissions from equipment used in the extraction process. Modern UK and EU regulation requires zero emissions of methane, a potent greenhouse gas. Escape of methane is a bigger problem in older wells than in ones built under more recent EU legislation.

Hydraulic fracturing uses between 1.2 and 3.5 million US gallons (4,500 and 13,200 m^3) of water per well, with large projects using up to 5 million US gallons (19,000 m^3). Additional water is used when wells are refractured. An average well requires 3 to 8 million US gallons (11,000 to 30,000 m^3) of water over its lifetime. According to the Oxford Institute for Energy Studies, greater volumes of fracturing fluids are required in Europe, where the shale depths average 1.5 times greater than in the U.S.Surface water may be contaminated through spillage and improperly built and maintained waste pits, and ground water can be contaminated if the fluid is able to escape the formation being fractured (through, for example, abandoned wells) or by produced water (the returning fluids, which also contain dissolved constituents such as minerals and brine waters). Produced water is managed by underground injection, municipal and commercialwastewater treatment and discharge, self-contained systems at well sites or fields, and recycling to fracture future wells. Typically less than half of the produced water used to fracture the formation is recovered.

About 3.6 hectares (8.9 acres) of land is needed per each drill pad for surface installations. Well pad and supporting structure construction significantly fragments landscapes which likely has negative effects on wildlife. These sites need to be remediated after wells are exhausted. Each well pad (in average 10 wells per pad) needs during preparatory and hydraulic fracturing process about 800 to 2,500 days of noisy activity, which affect both residents and local wildlife. In addition, noise is created by continuous truck traffic (sand, etc.) needed in hydraulic fracturing. Research is underway to determine if human health has been affected by air and water pollution, and rigorous following of safety procedures and regulation is required to avoid harm and to manage the risk of accidents that could cause harm.

In July 2013, the US Federal Railroad Administration listed oil contamination by hydraulic fracturing chemicals as "a possible cause" of corrosion in oil tank cars.

Hydraulic fracturing sometimes causes induced seismicity or earthquakes. The magnitude of these events is usually too small to be detected at the surface, although tremors attributed to fluid injection into disposal wells have been large enough to have often been felt by people, and to have caused property damage and possibly injuries.

Microseismic events are often used to map the horizontal and vertical extent of the fracturing. A better understanding of the geology of the area being fracked and used for injection wells can be helpful in mitigating the potential for significant seismic events.

Regulations

Countries using or considering use of hydraulic fracturing have implemented different regulations, including developing federal and regional legislation, and local zoning limitations. In 2011, after public pressure France became the first nation to ban hydraulic fracturing, based on the precautionary principle as well as the principle of preventive and corrective action of environmental hazards. The ban was upheld by an October 2013 ruling of the Constitutional Council. Some other countries such as Scotland have placed a temporary moratorium on the practice due to public health concerns and strong public opposition. Countries like the United Kingdom and South Africa have lifted their bans, choosing to focus on regulation instead of outright prohibition. Germany has announced draft regulations that would allow using hydraulic fracturing for the exploitation of shale gas deposits with the exception of wetland areas. In China, regulation on shale gas still faces hurdles, as it has complex interrelations with other regulatory regimes, especially trade.

The European Union has adopted a recommendation for minimum principles for using high-volume hydraulic fracturing. Its regulatory regime requires full disclosure of all additives. In the United States, the Ground Water Protection Council launched FracFocus.org, an online voluntary disclosure database for hydraulic fracturing fluids funded by oil and gas trade groups and the

U.S. Department of Energy. Hydraulic fracturing is excluded from the Safe Drinking Water Act's underground injection control's regulation, except when diesel fuel is used. The EPA assures surveillance of the issuance of drilling permits when diesel fuel is employed.

In 2012, Vermont became the first state in the United States to ban hydraulic fracturing. On 17 December 2014, New York became the second state to issue a complete ban on any hydraulic fracturing due to potential risks to human health and the environment.

Waterless Fracturing

Waterless fracturing is an alternative to hydraulic fracturing in which liquefied petroleum gas (LPG) which uses propane liquefied into gel. Propane is pumped into shale rock formation instead of water. Using propane can be considered more beneficial than water, because it does not block all pathways and therefore more natural gas is released. In addition, it does not carry poisonous chemicals and underground radioactivity back to the surface, and when it does comes back to the surface it can have another usage or be used again for LPG. The only drawback is the cost as propane is more expensive than water. Another drawback is that the possible leak of propane may lead to flash fire, so fracturing with propane requires more safety instructions and monitoring equipments in order to reduce possible risks.

According to Nathan Janiczek, the new era of waterless fracturing is growing and it is called Liquefied Petroleum Gas, or LPG fracturing. LPG is always pumped in a well and because of that pumping, rocks are destroyed and gas is released. Besides that, LPG is fully converted into gas when it is pumped up to the surface and that guarantees a one hundred percent retrieval rate. If we compare conventional fracking and fracking using propane gel, LPG fracking does not produce waste, nearly 100% of propane gas is pumped back while 50% of hydraulic frack fluid remains underground. It has lower viscosity, less surface tension and lower specific gravity, which makes the fluid more effective in transport and less dependent on pressure. LPG has many advantages; however, it is a new technique, which needs further research. Liquefied petroleum gas fracturing was developed by a GasFrac energy company, based in Calgary, Alberta. The Chief Technology Officer of the GasFrac company said that it is being used since 2008 in gas wells of Canada in Alberta, British Columbia, New Brunswick, Texas, Pennsylvania, Colorado, Oklahoma and New Mexico.

Hydraulic Fracturing Proppants

A proppant is a solid material, typically sand, treated sand or man-made ceramic materials, designed to keep an induced hydraulic fracture open, during or following a fracturing treatment. It is added to a *fracking fluid* which may vary in composition depending on the type of fracturing used, and can be gel, foam or slickwater–based. In addition, there may be unconventional fracking fluids. Fluids make tradeoffs in such material properties as viscosity, where more viscous fluids can carry more concentrated proppant; the energy or pressure demands to maintain a certain flux pump rate (flow velocity) that will conduct the proppant appropriately; pH, various rheological factors, among others. In addition, fluids may be used in low-volume well stimulation of high-permeability sandstone wells (20k to 80k gallons per well) to the high-volume operations such as shale gas and tight gas that use millions of gallons of water per well.

Conventional wisdom has often vacillated about the relative superiority of gel, foam and slickwater fluids with respect to each other, which is in turn related to proppant choice. For example, Zuber, Kuskraa and Sawyer (1988) found that gel-based fluids seemed to achieve the best results for coal-bed methane operations, but as of 2012, slickwater treatments are more popular.

Other than proppant, slickwater fracturing fluids are mostly water, generally 99% or more by volume, but gel-based fluids can see polymers and surfactants comprising as much as 7 vol% , ignoring other additives. Other common additives include hydrochloric acid (low pH can etch certain rocks, dissolving limestone for instance), friction reducers, guar gum, biocides, emulsion breakers, emulsifiers, 2-butoxyethanol, and radioactive tracer isotopes.

Proppant Permeability and Mesh Size

Sand used for fracturing, USGS, 2012.

Proppants used should be permeable or permittive to gas under high pressures; the interstitial space between particles should be sufficiently large, yet have the mechanical strength to withstand closure stresses to hold fractures open after the fracturing pressure is withdrawn. Large mesh proppants have greater permeability than small mesh proppants at low closure stresses, but will mechanically fail (i.e. get crushed) and produce very fine particulates ("fines") at high closure stresses such that smaller-mesh proppants overtake large-mesh proppants in permeability after a certain threshold stress.

Though sand is a common proppant, untreated sand is prone to significant fines generation; fines generation is often measured in wt% of initial feed. A commercial newsletter from Momentive cites untreated sand fines production to be 23.9% compared with 8.2% for lightweight ceramic and 0.5% for their product. One way to maintain an ideal mesh size (i.e. permeability) while having sufficient strength is to choose proppants of sufficient strength; sand might be coated with resin,- to form CRCS (Curable Resin Coated Sand) or PRCS (Pre-Cured Resin Coated Sands). In certain situations a different proppant material might be chosen altogether—popular alternatives include ceramics and sintered bauxite.

Proppant Weight and Strength

Increased strength often comes at a cost of increased density, which in turn demands higher flow rates, viscosities or pressures during fracturing, which translates to increased fracturing costs, both environmentally and economically. Lightweight proppants conversely are designed to be lighter than sand (~2.5 g/cm^3) and thus allow pumping at lower pressures or fluid velocities. Light proppants are less likely to settle. Porous materials can break the strength-density trend, or even afford greater gas permeability. Proppant geometry is also important; certain shapes or forms amplify stress on proppant particles making them especially vulnerable to crushing (a sharp discontinuity can classically allow infinite stresses in linear elastic materials).

Proppant Deposition and Post-treatment Behaviours

Proppant mesh size also affects fracture length: proppants can be "bridged out" if the fracture width decreases to less than twice the size of the diameter of the proppant. As proppants are deposited in a fracture, proppants can resist further fluid flow or the flow of other proppants, inhibiting further growth of the fracture. In addition, closure stresses (once external fluid pressure is released) may cause proppants to reorganise or "squeeze out" proppants, even if no fines are generated, resulting in smaller effective width of the fracture and decreased permeability. Some companies try to cause weak bonding at rest between proppant particles in order to prevent such reorganisation. The modelling of fluid dynamics and rheology of fracturing fluid and its carried proppants is a subject of active research by the industry.

Proppant Costs

Though good proppant choice positively impacts output rate and overall ultimate recovery of a well, commercial proppants are also constrained by cost. Transport costs from supplier to site form a significant component of the cost of proppants.

Other Components of Fracturing Fluids

Other than proppant, slickwater fracturing fluids are mostly water, generally 99% or more by volume, but gel-based fluids can see polymers and surfactants comprising as much as 7 vol% , ignoring other additives. Other common additives include hydrochloric acid (low pH can etch certain rocks, dissolving limestone for instance), friction reducers, guar gum,biocides, emulsion breakers, emulsifiers, and 2-Butoxyethanol.

Radioactive tracer isotopes are sometimes included in the hydrofracturing fluid to determine the injection profile and location of fractures created by hydraulic fracturing. Patents describe in detail how several tracers are typically used in the same well. Wells are hydraulically fractured in different stages. Tracers with different half-lives are used for each stage. Their half-lives range from 40.2 hours (lanthanum-140) to 5.27 years (cobalt-60). Amounts per injection of radionuclide are listed in The US Nuclear Regulatory Commission (NRC) guidelines. The NRC guidelines also list a wide range or radioactive materials in solid, liquid and gaseous forms that are used as field flood or enhanced oil and gas recovery study applications tracers used in single and multiple wells.

In the US, except for diesel-based additive fracturing fluids, noted by the American Environmental Protection Agency to have a higher proportion of volatile organic compounds and carcinogenic BTEX, use of fracturing fluids in hydraulic fracturing operations was explicitly excluded from regulation under the American Clean Water Act in 2005, a legislative move that has since attracted controversy for being the product of special interests lobbying.

Regulation of Hydraulic Fracturing

Countries using or considering to use hydraulic fracturing have implemented different regulations, including developing federal and regional legislation, and local zoning limitations. In 2011, after public pressure France became the first nation to ban hydraulic fracturing, based on the precautionary principle as well as the principal of preventive and corrective action of environmental hazards. The ban was upheld by an October 2013 ruling of the Constitutional Council. Some other countries have placed a temporary moratorium on the practice. Countries like the United Kingdom and South Africa, have lifted their bans, choosing to focus on regulation instead of outright prohibition. Germany has announced draft regulations that would allow using hydraulic fracturing for the exploitation of shale gas deposits with the exception of wetland areas.

The European Union has adopted a recommendation for minimum principles for using high-volume hydraulic fracturing. Its regulatory regime requires full disclosure of all additives. In the United States, the Ground Water Protection Council launched FracFocus.org, an online voluntary disclosure database for hydraulic fracturing fluids funded by oil and gas trade groups and the U.S. Department of Energy. Hydraulic fracturing is excluded from the Safe Drinking Water Act's underground injection control's regulation, except when diesel fuel is used. The EPA assures surveillance of the issuance of drilling permits when diesel fuel is employed.

On 17 December 2014, New York state issued a statewide ban on hydraulic fracturing, becoming the second state in the United States to issue such a ban after Vermont.

Approaches

Risk-based Approach

The main tool used by this approach is risk assessment. A risk assessment method, based on experimenting and assessing risk ex-post, once the technology is in place. In the context of hydraulic fracturing, it means that drilling permits are issued and exploitation conducted before the potential risks on the environment and human health are known. The risk-based approach mainly relies on a discourse that sacralizes technological innovations as an intrinsic good, and the analysis of such innovations, such as hydraulic fracturing, is made on a sole cost-benefit framework, which does not allow prevention or ex-ante debates on the use of the technology. This is also referred to as "learning-by-doing". A risk assessment method has for instance led to regulations that exist in the hydraulic fracturing in the United States (EPA will release its study on the effect of hydraulic fracturing on groundwater in 2014, though hydraulic fracturing has been used for more than 60 years. Commissions that have been implemented in the US to regulate the use of hydraulic fracturing have been created after hydraulic fracturing had started in their area of regulation. This is for instance the case in the Marcellus shale area where three regulatory committees were implemented ex-post.

Academic scholars who have studied the perception of hydraulic fracturing in the North of England have raised two main critiques of this approach. Firstly, it takes scientific issues out of the public debate since there is no debate on the use of a technology but on its effects. Secondly, it does not prevent environmental harm from happening since risks are taken then assessed instead of evaluated then taken as it would be the case with a precautionary approach to scientific debates. The relevance and reliability of risk assessments in hydraulic fracturing communities has also been debated amongst environmental groups, health scientists, and industry leaders. A study has epitomized this point: the participants to regulatory committees of the Marcellus shale have, for a majority, raised concerns about public health although nobody in these regulatory committees had expertise in public health. That highlights a possible underestimation of public health risks due to hydraulic fracturing. Moreover, more than a quarter of the participants raised concerns about the neutrality of the regulatory committees given the important weigh of the hydraulic fracturing industry. The risks, to some like the participants of the Marcellus Shale regulatory committees, are overplayed and the current research is insufficient in showing the link between hydraulic fracturing and adverse health effects, while to others like local environmental groups the risks are obvious and risk assessment is underfunded.

Precaution-based Approach

The second approach relies on the precautionary principle and the principal of preventive and corrective action of environmental hazards, using the best available techniques with an acceptable economic cost to insure the protection, the valuation, the restoration, management of spaces, resources and natural environments, of animal and vegetal species, of ecological diversity and equilibriums. The precautionary approach has led to regulations as implemented in France and Vermont, banning hydraulic fracturing.

Such an approach is called upon by social sciences and the public as studies have shown in the North of England and Australia. Indeed, in Australia, the anthropologist who studied the use of hydraulic fracturing concluded that the risk-based approach was closing down the debate on the ethics of such a practice, therefore avoiding questions on broader concerns that merely the risks implied by hydraulic fracturing. In the North of England, levels of concerns registered in the deliberative focus groups studied were higher regarding the framing of the debate, meaning the fact that people did not have a voice in the energetic choices that were made, including the use of hydraulic fracturing. Concerns relative to risks of seismicity and health issues were also important to the public, but less than this. A reason for that is that being withdrawn the right to participate in the decision-making triggered opposition of both supporters and opponents of hydraulic fracturing.

The points made to defend such an approach often relate to climate change and the impact on the direct environment; related to public concerns on the rural landscape for instance in the UK. Energetic choices indeed affect climate change since greenhouse gas emissions from fossil fuels extraction such as shale gas and oil contribute to climate change. Therefore, people have in the UK raised concerns about the exploitation of these resources, not just hydraulic fracturing as a method. They would hence prefer a precaution-based approach to decide whether or not, regarding the issue of climate change, they want to exploit shale gas and oil.

Framing of the Debate

There are two main areas of interest regarding how debates on hydraulic fracturing for the exploitation of unconventional oil and gas have been conducted.

"Learning-by-doing" and the Displacement of Ethics

A risk-based approach is often referred to as "learning-by-doing" by social sciences. Social sciences have raised two main critiques of this approach. Firstly, it takes scientific issues out of the public debate since there is no debate on the use of a technology but on its impacts. Secondly, it does not prevent environmental harm from happening since risks are taken then assessed instead of evaluated then taken. Public concerns are shown to be really linked to these issues of scientific approach. Indeed, the public in the North of England for instance fears "the denial of the deliberation of the values embedded in the development and application of that technology, as well as the future it is working towards" more than risks themselves. The legitimacy of the method is only questioned after its implementation, not before. This vision separates risks and effects from the values entitled by a technology. For instance, hydraulic fracturing entitles a transitional fuel for its supporters whereas for its opponents it represents a fossil fuel exacerbating the greenhouse effect and global warming. Not asking these questions leads to seeing only the mere economic cost-benefit analysis.

This is linked to a pattern of preventing non-experts from taking part in scientific-technological debates, including their ethical issues. An answer to that problem is seen to be increased public participation so as to have the public deciding which issues to address and what political and ethical norms to adopt as a society. Another public concern with the "learning-by-doing" approach is that the speed of innovation may exceed the speed of regulation and since innovation is seen as serving private interests, potentially at the expense of social good, it is a matter of public concern. Science and Technology Studies have theorized "slowing-down" and the precautionary principle as answers. The claim is that the possibility of an issue is legitimate and should be taken into account before any action is taken.

Variations in Risk-assessment of Environmental Effects of Hydraulic Fracturing

Issues also exist regarding the way risk assessment is conducted and whether it reflects some interests more than others. Firstly, an issue exists about whether risk assessment authorities are able to judge the impact of hydraulic fracturing in public health. A study conducted on the advisory committees of the Marcellus Shale gas area has shown that not a single member of these committees had public health expertise and that some concern existed about whether the commissions were not biased in their composition. Indeed, among 51 members of the committees, there is no evidence that a single one has any expertise in environmental public health, even after enlarging the category of experts to "include medical and health professionals who could be presumed to have some health background related to environmental health, however minimal". This cannot be explained by the purpose of the committee since all three executive orders of the different committees mentioned environmental public health related issues. Another finding of the authors is that a quarter of the opposed comments mentioned the possibility of bias in favor of gas industries in the composition of committees. The authors conclude saying that political leaders may not want to raise public health concerns not to handicap further economic development due to hydraulic fracturing.

Secondly, the conditions to allow hydraulic fracturing are being increasingly strengthened due to the move from governmental agencies' authority over the issue to elected officials' authority over it. The Shale Gas Drilling Safety Review Act of 2014 issued in Maryland forbids the issuance of drilling permits until a high standard "risk assessment of public health and environmental hazards relating to hydraulic fracturing activities" is conducted for at least 18 months based on the Governor's executive order.

Institutional Discourse and the Public

A qualitative study using deliberative focus groups has been conducted in the North of England, where there is a big shale gas reservoir exploited by hydraulic fracturing. These group discussions reflect many concerns on the issue of the use of unconventional oil and gas. There is a concern about trust linked with a doubt on the ability or will of public authorities to work for the greater social good since private interests and profits of industrial companies are seen as corruptive powers. Alienation is also a concern since the feeling of a game rigged against the public rises due to "decision making being made on your behalf without being given the possibility to voice an opinion". Exploitation also arises since economic rationality that is seen as favoring short-termism is accused of seducing policy-makers and industry. Risk is accentuated by what is hydraulic fracturing as well as what is at stake, and "blind spots" of current knowledge as well as risk assessment analysis are accused of increasing the potentiality of negative outcomes. Uncertainty and ignorance are seen as too important in the issue of hydraulic fracturing and decisions are therefore perceived as rushed, which is why participants favored some form of precautionary approach. There is a major fear on the possible disconnection between the public's and the authorities' visions of what is a good choice for the good reasons.

It also appears that media coverage and institutional responses are widely inaccurate to answer public concerns. Indeed, institutional responses to public concerns are mostly inadequate since they focus on risk assessment and giving information to the public that is considered anxious because ignorant. But public concerns are much wider and it appears that public knowledge on hydraulic fracturing is rather good.

The hydraulic fracturing industry has lobbied for permissive regulation in Europe, the US federal government, and US states. On March 20, 2015 the rules for disclosing the chemicals used were tightened by the Obama administration. The new rules give companies involved 30 days from the beginning of an operation on federal land to disclose those chemicals.

Hydro-slotted Perforation

Hydro-slotting perforation technology is the process of opening the productive formation through the casing and cement sheath to produce the oil or gas product flow (intensification, stimulation). The process has been used for industrial drilling since 1980, and involves the use of an underground hydraulic slotting engine (tool, equipment). The technology helps to minimize compressive stress following drilling in the well-bore zone (which reduces the permeability in the zone).

Overview

Since ancient times, when there were the first coal mines, it was observed, that increasing the depth of the development the coal tunnel, under the action of overburden pressure, surrounding rocks become harder and little-permeable. To solve this problem they developed a cavern of a certain form in the rock. More modern mining geo-mechanics explain the reason for the occurrence of this effect in relation to drilling wells. During any drilling process in the well there is formed the annular compressive stress conditions around the wellbore zone. The deeper the well, the more overburden pressure, which means the greater the annular compressive stress conditions. On the rocks lying at depths of 3–5 km the compressive stresses may reach up-to 75–125 MPa. In the near-well zone, as a result of concentration these stresses increase and sometimes become equal to double 150–250 MPa. If the tectonic stresses is several times higher than stresses from the weight of rocks, the stresses in the near-well zone may be even greater.

Under the action of stress conditions and high overburden pressure occurs a significant reduction in permeability in the near wellbore zone, in some cases close to zero. Oil or gas flow can not penetrate to the well. Traditional methods of opening the productive layer formation (cumulative, jet perforation, sand jet perforation, abrasive jetting perforation and other similar methods) did not consider this complicated situation in the near-well zone and therefore was not the effective. Porous and fractured formations are subjected to compression, that deforms the rock mass and reduces its permeability. The greater the depth, the stronger the effect can be.

Hydro-slotting perforation is quite different from jet (hydro-jetting or sand-blast) perforation. The energy of working fluid, consisting from water (layer water) and sand (abrasive quartz sand) pressure in the hydraulic engine, is divided into two components: five percent of energy goes to the creation of smooth uniform rectilinear motion of the working rod with the perforator and nozzles (between two and six nozzles) without participation in the process the multimeter tubing or coil-tubing. Ninety-five percent of energy goes to the cutting of continued and geometrically correct deep slots (up to five feet deep and between three and five slots at the same time). Slot length is equal to the length of the working engine shaft, usually 1.64 feet (0.00050 km).

The hydro-slotting perforation process does not deform the casing, does not create cracks in the cement, and does not clog-up the borders in the formation.

The geometry and depth of the slots creates the conditions for occurrence of the effect of unloading the circular stress conditions in the near wellbore zone (from 50 to 100 percent) and accordingly the increase of permeability (up to 30 to 50 percent) in this zone. In addition to this it forms a large area of the penetration (31.5 square feet (2.93 m^2) area for one cut with two nozzles only), that provides a very good hydrodynamic connection of the productive layer with the well.

The cutting speed may be corrected with the temperature in the borehole, temperature of the working fluid, concentration, flow and pressure. (these components are enough to completely control the depth and length of the cut and thus forming the slots), to instantly cut through the steel casing, through the cement to delve into the productive formation and keep the jets in this state while moving along the borehole, keeping the same depth of cut. At the end of the cutting continuous slot process the engine is set up to the initial position and ready for the next cutting interval. The process of hydro-slotting perforation and the depth of cut is controlled by the working fluid supply, pressure and concentration. The equipment can be operated without lifting on the surface for 11–15 hours.

Hydro-slotting perforation is the ecologically safe, environmentally friendly and effective affordable method for intensifying the operation in oil, gas, injection and hydro-geological wells. Now this method is widely used in Azerbaijan, Brazil, China, Eastern and Western Siberia, Jordan, Kazakhstan, Komi Republic, North Caucasus, Russia, Udmurtia, Ukraine, Urals, Uzbekistan and Yemen. The first mention regarding the hydro-slotting perforation in America, was in 1987 at the oil and gas conference in Texas. The first use of hydro-slotting perforation in the United States dates back to 1996, when together with Shell E & P Technology Company, discovered two wells (Abrasive Hydro jet Technology in Albert Load, Michigan). After that the hydro-slotting perforation was highly appreciated by the Department of Geophysics at Stanford University and by Division of Shell Exploration and Production by Shell E&P Technology Company. Hydro-slotting perforation was used in California, Kansas, Michigan, Montana, Nebraska, New York, Pennsylvania, Texas and Wyoming states. In Canada it has been successfully applied in Saskatchewan.

General Concepts

For opening of any productive layer it is necessary to open the casing, cement sheath and productive layer formation. Geophysics and mining geo-mechanics dictated the next requirements:

- Zone of cement sheath should be opened completely and not have cracks (to prevent possible overflows of water);
- Productive layer formation should be opened to maximum and on the maximum depth. At the same time productive formation should not have clogging, plugging, grouting, occlusive and cinder borders to produce excellent hydrodynamic connections of the productive layer with the well. Encompassing unloading the circular stress conditions around the wellbore, formed as a result of drilling, and increasing the permeability (50–100 percent) in the near wellbore zone (as a consequence of the first).

In the early 1970s, the Ministry of Geology of the USSR placed the Government order to scientific research institutions of the Country for the solution of the annular stress conditions and increase the permeability problem in the drilling wells. It was necessary to create the technology of opening the productive layer formation taking into account of uploading the annular stress conditions and increase the permeability in the near wellbore zone. The work to study this problem were assigned to the Institute of Oceanology and VNIMI (St. Petersburg, Russia). During the study there was done hundreds of experiments and mathematical models. It was determined, that if creating a geometrically correct, extended slot, directed along the wellbore and perpendicular to it on the distance from around 0.7 inches (18 mm) to 3.5 feet (1.1 m), in the zone of around the wellbore, there occurs the unloading of the annular compressive stress conditions from 50 to 100 percent, that are redirected to the far plane of the surface of formed slot, parallel to the wellbore surface. At the same time the permeability in this zone increased 30 to 50 percent. The holes after cumulative, jet perforation, sand jet perforation, abrasive jetting perforation and other similar methods, do not give the effect. The spot perforation did not create a slot in the casing and did not reach the required (unloading effect) depth, because the reverse jet interfered with direct jet and the maximum depth of the hole could not exceed 0.65 feet. When perforation with the movement occurs, the direct jet does not intersect with the reverse jet and depth of cutting can be much more (up to five feet) which is known as the excavation effect. Later it was proven mathematically.

It was necessary to create a device, that could make the continuation, along the borehole, slots in the casing, cement and go further into the productive formation. The tests with the movement of the multimeter tubing were not been successful, showing it was impossible to create geometrically correct extended slots with moving tubing. It was necessary to create an apparatus, that created a movement of cutting jets by itself, independently from tubing and located on the end of the tubing, directly in the leveled area. independent movement of the cutting jets could only be done mechanically, electrically or hydraulically. After another six months of research and testing it was decided to use mechanics and hydraulics as the base. The first prototype of hydro-slotting perforation device was created in 1972. The technology of hydro-slotting perforation was never sold to anyone. The hydro-slotting perforation technology was transformed into the category of performance techniques (as the technique of conducting the drilling, cumulative perforation, hydraulic fracturing, logging, pumping and so on).

The finalization of the device (prototype) in the end of 1972 was tasked to the special laboratory of the Research Institute of Oceanology of PSU "Sevmorgeo". From the beginning of the work for the revision the existing device was carried out in two directions: hydro-slotting perforation and hydro-mechanical slotting perforation. The second variant differs from the first in that at the beginning the opening of the casing is produced with a circular saw, and then the rock eroded by working fluid (water and sand) jets. The works were done over three years. The work for improvements of the hydro-mechanical slotting perforation were terminated in the result of their further inexpedient. Firstly, it was not necessary to divide the process into two operations: cutting the casing with the circular saw and a further jet-slotting perforation, because the cutting of casing with jet-slotting perforation takes place in a matter of seconds. Second, the mechanism of the circular saw takes up a lot of space in the housing unit, it was impossible to use the energy of working fluid to full power for getting deep slots, the slots get small and not deep (not enough for occurs the unloading of the annular compressive stress conditions and increase the permeability in the near wellbore zone). The further project was focused for finishing the hydro-slotting perforation device only.

In 1975, the scientific research laboratory of the Research Institute of Oceanology of PSU Sevmorgeo completed the project to improve the prototype of hydro-slotting perforation tool and this tool has been able to operate independently of the tubing movement. The equipment was 16 feet (4.9 m) long, 4.02 inches (10.2 cm) OD, weight 300 pounds (140 kg) and stroke length of 0.5 feet (0.15 m) only, and it worked on the following principle: the energy of working fluid pressure was divided into two components. Part of energy was used for the motion creation for the working rod with the perforator and nozzles; the other part of the energy was use for the cutting process (creating the continued slots along the wellbore through the casing and cement into the productive formation). The form and depth of the slots allowed the device to perform its main task, unloading the annular stress conditions and increase the permeability. The first practical tests in the wells were successfully made at the end of 1975 on "Archeda" field (Volgograd, Russia).

Benefits

Ability to increase area of development.

- Very deep penetration from three to six feet.
- Vertical permeability.
- Porosity increases four to five times.

- Permeability increases 15 times.
- Drainage volume increases six times.

Ability to access reserves which are otherwise inaccessible.

- In reservoirs located in close proximity to water, gas, and oil contacts.
- In weakly permeable, tightly-cemented reservoirs.
- In missed layers, or in layers covered by two or more columns.

Gentle approach with the ability to repair well-bore damage.

- In carbonates, dislodges clay particles and fines.
- In sandstones, reduces sand mobility problems.
- In deep gas sands, relieves overpressure damage from mud weight systems.
- Does not crack casing or cement.
- Maintains hydraulic integrity with no detonation impacts.
- Redistributes stresses away from the near-well-bore zone.

Development

During the period from the date of the first prototype of hydro-slotting perforation tool to present day, the type and technological characteristics of the equipment was significantly improved. The modern underground hydro-slotting equipment represents the devices, capable to instantly cut through the steel casing, through the cement to delve into the productive formation and keep the jets in this state while moving along the borehole, keeping the same depth of cut. Hydro-slotting equipment made of special high-strength materials, 12 feet (3.7 m) long, 3.5 feet (110 cm) OD, weight 180 pounds (82 kg), cutting speed from the point of perforation to 0.7 inches (1.8 cm) per minute, working stroke length 1.65 feet (0.50 m) (4.92 feet (1.50 m) x 1.64 feet (0.50 m) x 1.97 inches (5.0 cm) each slot), depth of slots five feet, continued and geometrically correct slots, opening area 63 square feet (5.9 m^2) per cut with four nozzles, can apply streamlined perforators between two and six nozzles, unloading the annular stress conditions in the near wellbore zone 50 to 100 percent, and increase the permeability 30 to 50 percent. The continuous time without lifting to the surface is 11–15 hours (nozzles lifespan ~ 15 hours, perforator ~ seven wells, hydraulic engine ~ 40 wells).

Without lifting to the surface with the hydro-slotting tool can also:

- cut on the previous perforation (cumulative perforation).
- colmatation treatment.
- cut the thin-interbedded layers.
- mini hydraulic fracture stimulation.

- create the continuous slot.
- cut the shale.
- accurate cut near the water reservoir or opposite in the injection wells.
- bypass the water layers.
- bypass the casing collars.
- cut a few casings.
- chemical treatment.
- sealing, direct and reverse flushing.
- tubing pressure testing.
- cut the casing at abandonment.

The hydro-slotting perforation process does not deform the casing, does not create cracks in the cement and does not clog up the borders in the formation. The process of hydro-slotting perforation is controlled. The cutting speed and depth of cutting may be corrected with the temperature in the borehole, temperature of the working fluid, concentration, flow and pressure. At the end of the cutting process of a continuous slot the engine is set up to the initial position and ready for the next cutting interval. Hydro-slotting perforation sets the perfect geometry for the subsequent fracturing. Hydro-slotting perforation can be applied in any formation: shale, carbonates, sandstone and so on.

Further improvement of the equipment for hydro-slotting perforation must follow the scientific and technical progress in this technology, not on the way of mindless increase of the holes in the hydro jets pipe. It is necessary to make the underground hydraulic engine for horizontal wells, which must be sealed to prevent the ingress of sand and mud inside and maintain the centerline position relative to the wellbore. It is necessary to make a self-orientation perforator (a particularly important issue of orientation in horizontal wells). For the orientation of the tool it is necessary there is communication with the tool (preferably two-sided) and surface of the well. Taking into account the specific conditions of hydro-slotting perforation process, signaling from the tool and back possibly using ultrasound only. Then the cutting process can be fully controlled from the surface, and it will be possible to change the speed and depth of cutting the slots regardless of the temperature inside the well.

Patents

Over the years this method has not undergone much change, but there are many patents on the method of hydro-slotting perforation. With the development of technological progress there has been continuously improved and refined equipment, but patents, regarding the hydro-slotting equipment in full is not so much, there are a few patents on parts.

- United States patents for complete hydro-slotting perforation equipment: US 8240369 B1, US 31,084.

- Similar United States patents: US3130786, US4227582, US5337825, US6651741, US7073587, US7140429, US7568525, US20070187086, US20090101414, USRE21085, 166/55.2, 166/298, and E21B43/114.
- United States patent for method of hydro-slotting perforation: US 20130105163 A1.
- Similar United States patents: US3130786, US4047569, US4134453, US5445220, US6564868, US7568525, and US20050269100.

Uses of Radioactivity in Oil and Gas Wells

Radioactive sources are used for logging formation parameters. Radioactive tracers, along with the other substances in hydraulic-fracturing fluid, are sometimes used to determine the injection profile and location of fractures created by hydraulic fracturing.

Use of Radioactive Sources for Logging

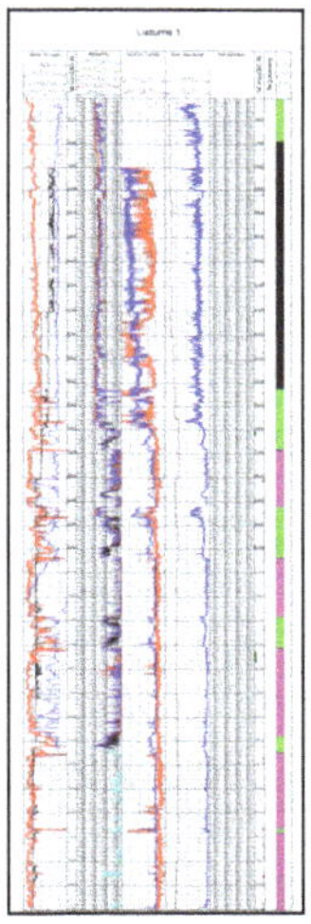

Composite wireline log for the Lisburne 1 well, Alaska - the neutron and density logs used radioactive sources.

Sealed radioactive sources are routinely used in formation evaluation of both hydraulically fractured and non-fracked wells. The sources are lowered into the borehole as part of the well logging tools, and are removed from the borehole before any hydraulic fracturing takes place. Measurement of formation density is made using a sealed caesium-137 source. This bombards the formation with high energy gamma rays. The attenuation of these gamma rays gives an accurate measure of formation density; this has been a standard oilfield tool since 1965. Another source is americium beryllium (Am-Be) neutron source used in evaluation of the porosity of the formation, and this has been used since 1950. In a drilling context, these sources are used by trained personnel, and radiation exposure of those personnel is monitored. Usage is covered by licenses from International Atomic Energy Agency (IAEA) guidelines, SU or European Union protocols, and the Environment Agency in the UK. Licenses are required for access, transport, and use of radioactive sources. These sources are very large, and the potential for their use in a 'dirty bomb' means security issues are considered as important. There is no risk to the public, or to water supplies under normal usage.

They are transported to a well site in shielded containers, which means exposure to the public is very low, much lower than the background radiation dose in one day.

Radiotracers and Markers

The oil and gas industry in general uses unsealed radioactive solids (powder and granular forms), liquids and gases to investigate or trace the movement of materials. The most common use of these radiotracers is at the well head for the measurement of flow rate for various purposes. A 1995 study found that radioactive tracers were used in over 15% of stimulated oil and gas wells.

Use of these radioactive tracers is strictly controlled. It is recommended that the radiotracer is chosen to have readily detectable radiation, appropriate chemical properties, and a half life and toxicity level that will minimize initial and residual contamination. Operators are to ensure that licensed material will be used, transported, stored, and disposed of in such a way that members of the public will not receive more than 1 mSv (100 mrem) in one year, and the dose in any unrestricted area will not exceed 0.02 mSv (2 mrem) in any one hour. They are required to secure stored licensed material from access, removal, or use by unauthorized personnel and control and maintain constant surveillance of licensed material when in use and not in storage. Federal and state nuclear regulatory agencies keep records of the radionuclides used.

As of 2003 the isotopes Antimony-124, argon-41, cobalt-60, iodine-131, iridium-192, lanthanum-140, manganese-56, scandium-46, sodium-24, silver-110m, technetium-99m, and xenon-133 were most commonly used by the oil and gas industry because they are easily identified and measured.Bromine-82, Carbon-14, hydrogen-3, iodine-125 are also used.

Examples of amounts used are:

Nuclide	Form	Activity
Iodine-131	Gas	100 millicuries (3.7 GBq) total, not to exceed 20 mCi (0.74 GBq) per injection
Iodine-131	Liquid	50 millicuries (1.9 GBq) total, not to exceed 10 mCi (0.37 GBq) per injection
Iridium-192	"Labeled" frac sand	200 millicuries (7.4 GBq) total, not to exceed 15 mCi (0.56 GBq) per injection
Silver-110m	Liquid	200 millicuries (7.4 GBq) total, not to exceed 10 mCi (0.37 GBq) per injection

In hydraulic fracturing, plastic pellets coated with Silver-110m or sand labelled with Iridium-192with may be added to a proppant when it is required to evaluate whether a fracturing process has penetrated rocks in the pay zone. Some radioactivity may by brought to the surface at the well head during testing to determine the injection profile and location of fractures. Typically this uses very small (50 kBq) Cobalt-60 sources and dilution factors are such that the activity concentrations will be very low in the topside plant and equipment.

Regulation in the US

The NRC and approved state agencies regulate the use of injected radionuclides in hydraulic fracturing in the United States.

The US EPA sets radioactivity standards for drinking water. Federal and state regulators do not require sewage treatment plants that accept gas well wastewater to test for radioactivity. In Pennsylvania, where the hydraulic fracturing drilling boom began in 2008, most drinking-water intake plants downstream from those sewage treatment plants have not tested for radioactivity since before 2006. The EPA has asked the Pennsylvania Department of Environmental Protection to require community water systems in certain locations, and centralized wastewater treatment facilities to conduct testing for radionuclides.

References

- Zukerman, Gregory (2013-11-06). "Breakthrough: The Accidental Discovery That Revolutionized American Energy". The Atlantis. Retrieved 18, March 2020
- Bell, Frederic Gladstone (2004). Engineering Geology and Construction. Taylor & Francis. p. 670. ISBN 9780415259392
- Gidley, John L. (1989). Recent Advances in Hydraulic Fracturing. SPE Monograph. 12. SPE. p. ?. ISBN 9781555630201
- Zoback, M.D. (2007). Reservoir geomechanics. Cambridge University Press. p. 18. ISBN 9780521146197. Retrieved 6, March 2012
- Gill, R. (2010). Igneous rocks and processes: a practical guide. John Wiley and Sons. p. 102. ISBN 978-1-4443-3065-6. Retrieved 11, August 2020
- Mader, Detlef (1989). Hydraulic Proppant Fracturing and Gravel Packing. Elsevier. pp. 173–174; 202. ISBN 9780444873521
- Miller, Bruce G. (2005). Coal Energy Systems. Sustainable World Series. Academic Press. p. 380. ISBN 9780124974517
- Chilingar, George V.; Robertson, John O.; Kumar, Sanjay (1989). Surface Operations in Petroleum Production. 2. Elsevier. pp. 143–152. ISBN 9780444426772
- Wan Renpu (2011). Advanced Well Completion Engineering. Gulf Professional Publishing. p. 424. ISBN 9780123858689
- Price, N. J.; Cosgrove, J. W. (1990). Analysis of geological structures. Cambridge University Press. pp. 30–33. ISBN 978-0-521-31958-4. Retrieved 5, August 2020
- Ching H. Yew (1997). Mechanics of Hydraulic Fracturing. Gulf Professional Publishing. p. ?. ISBN 9780884154747
- Brown, Edwin Thomas (2007) [2003]. Block Caving Geomechanics (2nd ed.). Indooroopilly, Queensland: Julius Kruttschnitt Mineral Research Centre, UQ. ISBN 978-0-9803622-0-6. Retrieved 14, May 2020
- Fjaer, E. (2008). "Mechanics of hydraulic fracturing". Petroleum related rock mechanics. Developments in petroleum science (2nd ed.). Elsevier. p. 369. ISBN 978-0-444-50260-5. Retrieved 14, July 2020
- Gold, Russell (2014). The Boom: How Fracking Ignited the American Energy Revolution and Changed th World. New York: Simon & Schuster. pp. 115–121. ISBN 978-1-4516-9228-0
- Charlez, Philippe A. (1997). Rock Mechanics: Petroleum Applications. Paris: Editions Technip. p. 239. ISBN 9782710805861. Retrieved 14, May 2020
- "New Waterless Fracking Method Avoids Pollution Problems, But Drillers Slow to Embrace It". insideclimatenews.org. Retrieved 05, July 2020

5

Diverse Types of Enhanced Oil Recovery

The recovery of hydrocarbons refers to the extraction of crude oil and natural gases. It involves various steps; primary recovery, secondary recovery, tertiary recovery and infill recovery. Primary and secondary recovery processes accounts for the 20-40% recovery of oils while tertiary recovery can extract 30-60% more reservoir oil. This chapter discusses in detail the theories related to enhanced oil recovery.

Recovery of Hydrocarbons

The recovery of hydrocarbons is basically a volume displacement process. When a volume of hydrocarbon is removed from the reservoir by production, it will be replaced by a volume of some fluid. Energy is expended in this process. Hydrocarbon recovery mechanisms may be divided into two categories:

(a) Primary Recovery.

(b) Enhanced Recovery.

Primary Recovery

Primary recovery is utilization of the natural energy of the reservoir to cause the hydrocarbon to flow into the wellbore. Based on this definition, as long as the hydrocarbon flows into the wellbore, this is primary recovery, even if the hydrocarbon must be artificially lifted to the surface by pumps or some other process. There are many sources of this primary recovery energy of which three are dominant:

(a) Dissolved Gas Drive (Solution Gas Drive).

(b) Gas-Cap Drive.

(c) Water Drive.

Dissolved Gas Drive

When the reservoir is produced so that gas is permitted to escape from the hydrocarbon liquid in the reservoir, so that two-phase flow (gas and liquid) occurs from the reservoir into the wellbore, the expanding gas will force the oil ahead of the gas into the wellbore.

In order to maximize oil recovery, however, for most reservoirs it is desirable to prevent dissolved gas drive, at least until late in the productive life of the reservoir. As the reservoir approaches depletion, the flowing bottomhole pressures may be reduced to as low a value as possible, in order to recover whatever percentage of remaining hydrocarbons might flow into the wellbore, including solution gas from the oil which will remain in the reservoir (residual oil) at the time the reservoir is abandoned.

Dissolved gas drive can be delayed by injecting water into the water zone beneath the oil, or gas on top of the oil there creating a gas cap, in order to maintain reservoir fluid pressures above the bubble point pressure.

Gas Cap Drive

If a gas cap exists above the oil zone, and wells are drilled and perforated in the oil zone and the bottomhole pressures are sufficiently reduced, the expanding gas cap will force the oil into the wells as the gas interface encroaches into the oil zone. In order for gas-cap drive to exist as a primary recovery mechanism, however, the gas cap must exist naturally.

Water Drive

Most hydrocarbon reservoirs will have a water zone beneath the hydrocarbon. This water is tending to encroach into the oil zone. If wells are drilled and perforated in the oil zone, when the wellbore pressure is reduced, oil flow will be initiated into the well as water encroaches into the oil zone forcing the oil towards the producing wells. If this natural encroachment tendency is to exist, natural energy must be present. There are several possible sources of this natural energy. One source is the expansion of the water as a compressible fluid, as reservoir pressures are reduced. As the reservoir pressure is reduced, the expanding water will push the oil in front of it into the producing wells. Water expansion as a compressed liquid produces more oil than oil as a compressed liquid, not because the compressibility of water is much different to compressibility of oil, but because the total volume of water in the water zone is usually very large when compared to the total volume of oil in the oil zone.

Another source of energy for water drive occurs when the reservoir rock dips upward to the surface where it outcrops. If permeability continuity exists through this rock, as oil is produced from the reservoir, water flows down dip from the surface to replace the oil volume removed. Surface water replenished that water, maintaining a constant hydrostatic pressure on the reservoir fluids.

Secondary Recovery

Secondary recovery is proven technology; indeed, a recent study indicates that 50 percent of all domestic crude oil in the US comes from secondary recovery operations.

Water flooding is inherently more efficient than gas displacement in pressure-maintenance projects and is the preferred process where feasible. Some reservoirs, principally those containing heavy oil that flows only with great difficulty, not only provide poor primary recovery but often are not susceptible to waterflooding. Enhanced oil recovery would be especially useful in some of these reservoirs.

Water Flood

Of the historical techniques used for EOR, water flooding has been the most common. This is not water drive. In water drive, water is encroaching into the oil zone from beneath, but in a true water flood, water is injected down injection wells into the oil zone. Ideally, this creates a vertical flood front, pushing the oil in front of the water toward the producing wells. In a water flood, the water injection wells are placed relative to the oil producing wells in some predetermined pattern based on reservoir characteristics and production history. A common pattern for water flooding for large reservoirs which arc basically horizontal reservoirs is the five spot patterns. This five spot pattern is repeated over the reservoir.

Prior to the initiation of a water flood project for a reservoir, various studies will have been made in designing the water flood. These might include model studies in the laboratory, digital and analog computer simulations, and pilot floods may have been run in a portion of the reservoir as a preliminary study, so that an analysis of the water flood plan might be made.

It is desirable to conduct the water flood so as to maximize the sweep efficiency within economic limits relative to production, so that when the water front from the injection wells breaks into the producing wells, a maximum percent of the reservoir volume will have been swept by the flood. Once this water front reaches the producing wells, further hydrocarbon production will be negligible, in that the wells will now produce essentially water. In order to recover further hydrocarbons, a different EOR technique must now be applied as a tertiary (or third) method for recovery.

Whatever the technique used for enhanced recovery, it is desirable that the mobility ratio of driving fluid be less than the mobility ratio for the driven fluid. The mobility ratio is the ratio of the permeability to the flow of the liquid to the dynamic viscosity of that liquid. The oil ratio mobility ratio will be;

[ko/mo] = Oil Mobility Ratio

And, in the case of the water flood, the water mobility ratio of the water will be;

[kw/mw] = Water Mobility Ratio

If the mobility ratio of the driving fluid is greater than the mobility ratio of the driven fluid, the driving fluid will tend to channel or finger through the hydrocarbon, tending to bypass the hydrocarbon in the smaller permeability channels, leaving it behind in the reservoir.

Gas Cap Injection

In the gas cap drive injection secondary recovery technique, gas is injected into the gas cap above the oil zone, to pressurize the gas cap. In reservoirs where reservoir fluid pressure is higher than the bubble point pressure, a gas cap may be created by gas injection so that the expending gas cap with further gas injection will displace the oil into the producing wells. As previously discussed, gas cap drive or gas cap drive enhancement is often used as a reservoir pressure maintenance technique.

Enhanced Recovery

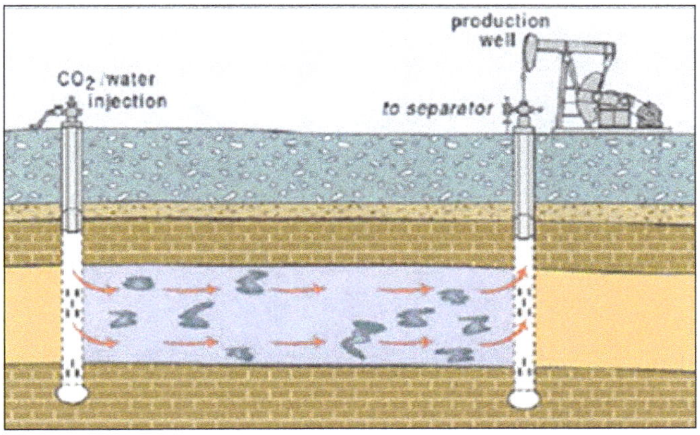

Processes that inject fluids other than natural gas and water to augment a reservoir's ability to produce oil have been designated "improved," "tertiary," and "enhanced" oil recovery processes. The term used in this assessment is enhanced oil recovery (EOR). According to American Petroleum Institute estimates of original oil in place and ultimate recovery, approximately two-thirds of the oil discovered will remain in an average reservoir after primary and secondary production. This inefficiency of oil recovery processes has long been known and the knowledge has stimulated laboratory and field testing of new processes for more than 50 years.

Early experiments with un-conventional fluids to improve oil recovery involved the use of steam and air for combustion to create heat. Current EOR processes may be divided into four categories:

(a) Thermal,

(b) Miscible,

(c) Chemical,

(d) Other.

Most EOR processes represent essentially untried, high-risk technology. One thermal process has achieved moderately widespread commercialization. The mechanisms of miscible processes are reasonably well understood, but it is still difficult to predict whether they will work and be profitable in any given reservoir. The chemical processes are the most technically complex, but they also could produce the highest recovery efficiencies.

The potential applicability of all EOR processes is limited not only by technological constraints, but by economic, material, and institutional constraints as well.

Thermal Processes

Viscosity, a measure of a liquid's ability to flow, varies widely among crude oils. Some crudes flow like road tar, others as readily as water. High viscosity makes oil difficult to recover with primary or secondary production methods.

The viscosity of most oils dramatically decreases as temperature increases, and the purpose of all thermal oil-recovery processes is therefore to heat the oil to make it flow or make it easier to drive with injected fluids. An injected fluid may be steam or hot water (steam injection), or air (combustion processes).

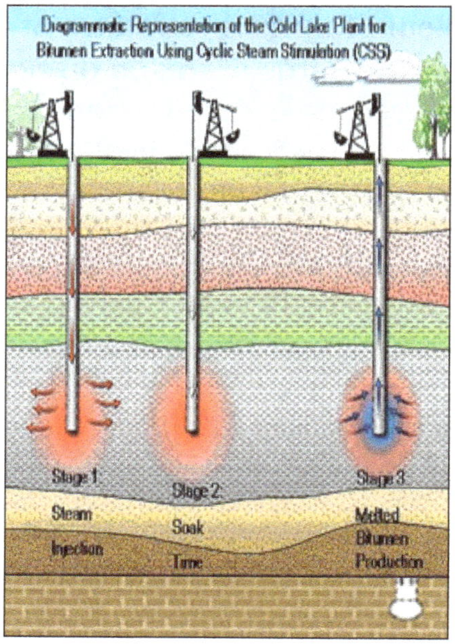

Steam Injection

Steam injection is the most advanced and most widely used EOR process. It has been successfully used in some reservoirs in California since the mid-1960's. There are two versions of the process: cyclic steam and steam drive. In the first, high-pressure steam or steam and hot water is injected into a well for a period of days or weeks. The injection is stopped and the reservoir is allowed to "soak." After a few days or weeks, the well is allowed to backflow to the surface. Pressure in the producing well is allowed to decrease and some of the water that condensed from steam during injection or that was injected as hot water then vaporizes and drives heated oil toward the producing well.

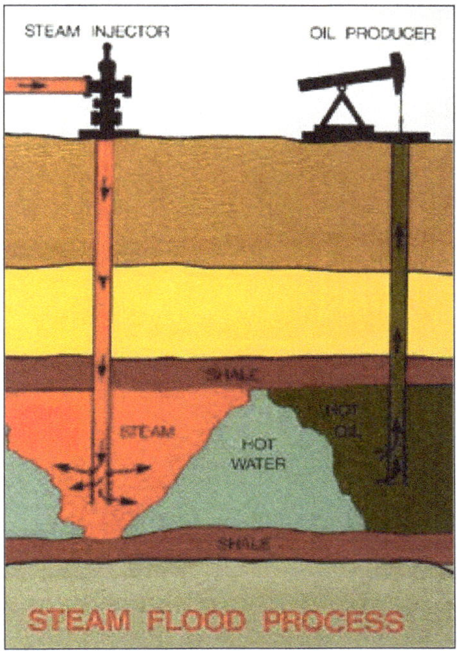

When oil production has declined appreciably, the process is repeated. Because of its cyclic nature, this process is occasionally referred to as the "huff and puff" method. The second method, steam drive or steam flooding, involves continuous injection of steam or steam and hot water in much the same way that water is injected in water flooding. A reservoir or a portion thereof is developed with interlocking patterns of injection and production wells. During this process, a series of zones develop as the fluids move from injection well to producing well. Nearest the injection well is a steam zone, ahead of this is a zone of steam condensate (water), and in front of the condensed water is a band or region of oil being moved by the water. The steam and hot water zone together remove the oil and force it ahead of the water.

Cyclic steam injection is usually attempted in a reservoir before a full-scale steam drive is initiated, partially as a means of determining the technical feasibility of the process for a particular reservoir and partly to improve the efficiency of the subsequent steam drive. A steam drive, where applicable, will recover more oil than cyclic steam injection.

Combustion Processes

Combustion projects are technologically complex, and difficult to predict and control. Injection of hot air will cause ignition of oil within a reservoir. Although some oil is lost by burning, the hot combustion product gases move ahead of the combustion zone to distill oil and push it toward producing wells. Air is injected through one pattern of wells and oil is produced from another interlocking pattern of wells in a manner similar to waterflooding. This process is referred to as fire flooding, in situ (in place) combustion, or forward combustion. Although originally conceived to apply to very viscous crude oils not susceptible to water flooding, the method is theoretically applicable to a relatively wide range of crude oils.

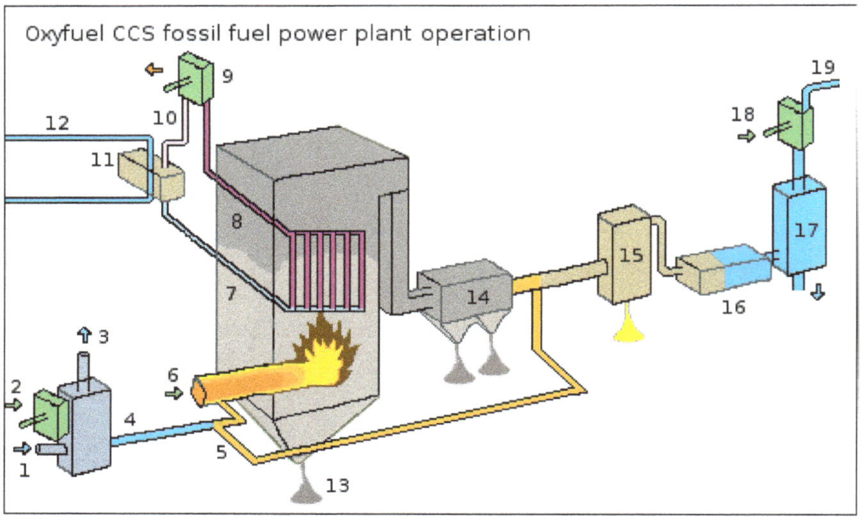

An important modification of forward combustion is the wet combustion process. Much of the heat generated in forward combustion is left behind the burning front. This heat was used to raise the temperature of the rock to the temperature of the combustion. Some of this heat may be recovered by injection of alternate slugs of water and air. The water is vaporized when it touches the hot formation. The vapor moves through the combustion zone heating the oil ahead of it and assists the production of oil.

Miscible Processes

Miscible processes are those in which an injected fluid dissolves in the oil it contacts, forming a single oil-like liquid that can flow through the reservoir more easily then the original crude. A variety of such processes have been developed using different fluids that can mix with oil, including alcohols, carbon dioxide, petroleum hydrocarbons such as propane or propane-butane mixtures, and petroleum gases rich in ethane, propane, butane, and pentane. The fluid must be carefully selected for each reservoir and type of crude to ensure that the oil and injected fluid will mix. The cost of the injected fluid is quite high in all known processes, and therefore either the process must include a supplementary operation torecover expensive injected fluid, or the injected material must be used sparingly. In this process, a "slug," which varies from 5 to 50 percent of the reservoir volume, is pushed through the reservoir by gas, water (brine), or chemically treated brine to contact and displace the mixture of fluid and oil.

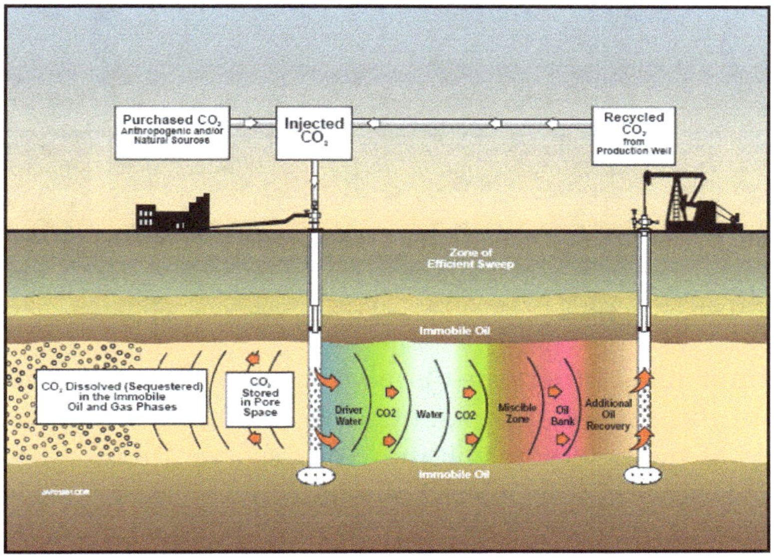

Miscible processes involve only moderately complex technology compared with other EOR processes. Although many miscible fluids have been field tested, much remains to be determined about the proper formulation of various chemical systems to affect complete solubility and to maintain this solubility in the reservoir as the solvent slug is pushed through it. Because of the high value of hydrocarbons and chemicals derived from hydrocarbons, it is generally felt that such materials would not make desirable injection fluids under current or future economic conditions. For this reason, attention has turned to Co2 as a solvent. Conditions for complete mixing of Co2 with crude oil depend on reservoir temperature and pressure and on the chemical nature and density of the oil.

Chemical Processes

Three EOR processes involve the use of chemicals: surfactant/polymer, polymer, and alkaline flooding.

Surfactant/Polymer Flooding

Surfactant/polymer flooding, also known as microemulsion flooding or micellar flooding, is the newest and most complex of the EOR processes. While it has a potential for superior oil recovery, few major field tests have been completed or evaluated. Several major tests are now under way to determine its technical and economic feasibility. Surfactant/polymer flooding can be any one of several processes in which detergent-like materials are injected as a slug of fluid to modify the chemical interaction of oil with its surroundings.

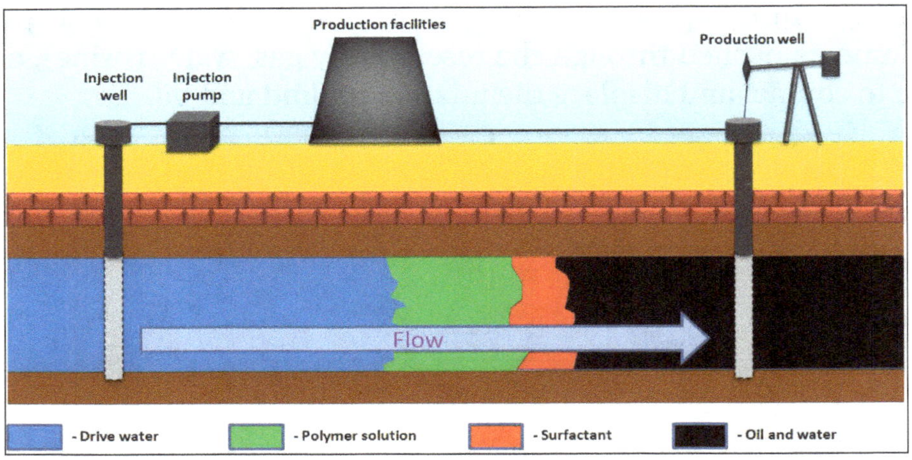

These processes emulsify or otherwise dissolve or partly dissolve the oil within the formation. Because of the cost of such agents, the volume of a slug can represent only a small percentage of the reservoir volume. To preserve the integrity of the slug as it moves through the reservoir, it is pushed by water to which a polymer has been added. The chemical composition of a slug and its size must be carefully selected for each reservoir/ crude oil system. Not all parameters for this design process are well understood.

Polymer Flooding

Polymer flooding is a chemically augmented waterflood in which small concentrations of chemicals, such as polyacrylamides or polysaccharides, are added to injected water to increase the effectiveness of the water in displacing oil.

Alkaline Flooding

Water solutions of certain chemicals such as sodium hydroxide, sodium silicate, and sodium carbonate are strongly alkaline. These solutions will react with constituents present in some crude oils or present at the rock/crude oil interface to form detergent-like materials which reduce the ability of the formation to retain the oil. The few tests which have been reported are technically encouraging, but the technology is not nearly so well developed as those described previously. Reservoirs not considered for alkaline flooding became candidates for other processes.

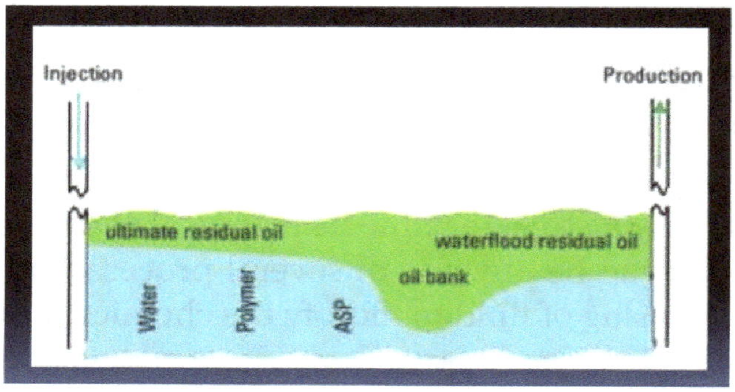

Other EOR Processes

Over the years, many processes for improving oil recovery have been developed, a large number of patents have been issued, and a significant number of processes have been field tested. In evaluating a conceptual process, it should be recognized that a single field test or patent represents but a small step toward commercial use on a scale large enough to influence the supply of crude oil. Some known processes have very limited application, for example, if thin coalbeds lay under an oil reservoir this coal could be ignited, the oil above it would be heated, its viscosity would be reduced, and it would be easier to recover. This relationship between oil and coal is rare, however, and the process is not important to total energy production. Another example involves use of electrical energy to fracture an oil-bearing formation and form a carbon track or band between wells. This band would then be used as a high resistance electrical pathway through which electric current would be applied, causing the "resistor" to heat the formation, reduce oil viscosity, and increase oil recovery. The process was conceived over 25 years ago and has been tested sporadically, but does not appear to have significant potential.

A third process in this category is the use of bacteria for recovery of oil. Several variations have been conceived. These include use of bacteria within a reservoir to generate surface-active (detergent-like) materials that would perform much the same function as a surfactant/polymer flood. Although some bacteria are able to with-stand temperatures and pressure found in oil reservoirs, none have been found that will both successfully generate useful modifying chemicals in sufficient amounts and also tolerate the chemical and thermal environments in most reservoirs. It is uncertain whether nutrients to keep them alive could be provided. Further, any strain of bacteria developed would need to be carefully screened for potential environmental hazards.

Recovery Efficiencies

Experience has determined expected ranges of efficiencies of recovery of hydrocarbons by primary and enhanced techniques. These recovery efficiencies are normally expressed in one of two ways:

(i.) Percent of Original-Oil-in-Place recovered.

(ii.) Percent of remaining-Oil-in-Place recovered.

The ranges of recovery efficiencies for primary recovery and enhanced may be summarized as follows:

- Primary Recovery Efficiencies

Oil (Percent of Original Oil- in- Place)

Dissolved Gas Drive	5% to 30%

Gas-Cap Drive	20% to 40%
Water Drive	35% to 75%

- Gas (Percent of Original-Gas-In –Place):

Gas Expansion and Water Drive	90% +

- Enhanced Recovery Efficiencies

Oil (Percent of Original- Oil- In- Place):

Water flood (Secondary Recovery)	30% to 40%
CO_2 Miscible Flood (Tertiary Recovery)	5% to 10%
Steam Drive (Heavy Oil)	50% to 80%

Primary Recovery

Primary recovery is the first stage of petroleum and gas production. Crude oil extraction from a new well relies on the natural rise of the oil due to pressure differences between the oil field and the bottom-hole of the well. Mechanical lift systems such as a rod pump are also a primary recovery method.

Primary recovery is also known as primary production. Primary recovery is less expensive than secondary and enhanced oil recovery (EOR). Enhanced oil recovery techniques are costly and use gases, chemicals, and heat to extract the oil. EOR is expensive and not always useful. Primary recovery takes advantage of the natural tendency for crude oil to rise to the surface once a well punctures the underground oil field.

The crude oil, contained in the ground, is under intense pressure, whereas the hollow well shaft is at lower pressure. Oil will flow rapidly into the area of lowest pressure into the well and up to the surface. When oil under pressure is uncontained, it can result in an oil geyser, spouting from the earth. During primary recovery typically only 5- to 15-percent of a well's total of potential hydrocarbons are extracted.

Hydrocarbons are organic chemical compounds, composed exclusively of hydrogen and carbon atoms. Hydrocarbons can be solids, liquids or gasses. Petroleum and natural gas are made primarily of hydrocarbons. These compounds combust in the presence of sufficient oxygen and produce carbon dioxide, water, and heat. Methane is the primary component of natural gas and is the simplest hydrocarbon because of its structured.

Oil and gas companies will use an estimated ultimate recovery (EUR) calculation to determine if the oil or gas contained in a field has the potential to make the hydrocarbon.

Natural Processes Affecting Primary Recovery

Various factors can cause the natural pressures that drive oil to the surface during primary recovery. One common primary recovery method is a gas drive. The gas drive uses the energy of expanding underground gas to force the oil to the surface. Another recovery method is the water drive. Water drives use underground aquifers to exert pressure on the oil. Also, in some shallow and steeply graded oil fields, the oil will drain to the surface with the force of gravity.

As production continues, the reservoir pressure will decrease, and hence the differential pressure will decrease. This decrease in pressure may necessitate the use of an artificial lifting system to continue production. The most common artificial lift for use in primary recovery is the rod pump. The rod pump employs a beam-and-crank assembly to create a reciprocating motion which transfers to vertical lift through a series of plungers and valves. This method is the classic oil derrick with its distinctive bobbing horse head.

Eventually, the primary recovery reaches its limit. This limit happens when the reservoir pressure is too low, or when the mix of gas or water into the output stream is too high. At this point, even artificial lift systems are not economical for the continued extraction of the hydrocarbons.

Solution Gas Drive

Solution gas drive is a mechanism by which dissolved gas in a reservoir will expand and become an energy support to produce reservoir fluid. Solution gas drive has other name, such as dissolved gas drive or depletion drive.

When reservoir pressure is more than the bubble point, no free gas presents in a reservoir and this is called "under saturated reservoir." At this stage, the drive comes from oil and connate water expansion and the compaction of reservoir pore space. Because compressibility of oil and rock is very low, only a small amount of fluid can be produced and typically the volume is around 1-2% of oil in place.

When reservoir pressure reaches a bubble point, oil becomes saturated and free gas will present in a reservoir. The expansion of gas is a main energy to produce reservoir fluid for the solution gas drive. At the beginning, the produced gas oil ratio will be slightly decline because free gas in a reservoir cannot move until it goes over the critical gas saturation. Then gas will begin to flow into a well. In some cases, where vertical permeability is high, gas may migrate up and become a secondary gas cap, which helps oil production.

When pressure gets lower, more gas will be produced and oil production will decline. This will lead to a high producing gas oil ratio. This is not a good sign because reservoir pressure declines sharply with gas production and eventually energy sources in a reservoir will drop and oil cannot be produced. Figure below shows general profiles of reservoir pressure, oil production, and Gas Oil Ratio (GOR) over a period of production.

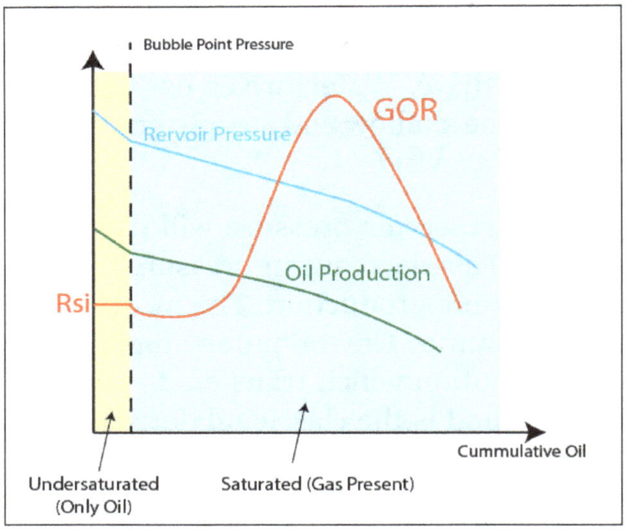

Solution Gas Drive.

This is very critical to perform the secondary recovery as water injection to maintain reservoir pressure above the bubble point so as to improve the oil recovery factor. Typical recovery factor from the solution gas drive reservoir is about 5 – 30%.

Gas Cap Drive

Hydrocarbons can experience equilibrium of phases, depending on the pressure and temperature conditions. In many cases, the vapour phase is less dense compared to the liquid, which leads to the accumulation of gas in the upper portions of the porous media forming a gas-cap. In the gas-cap mechanism, the energy comes from its expansion: as oil is produced, the gas-cap expands pushing the gas-oil contact downwards. As gases have high compressibilities, this expansion occurs without significant pressure decrease. Thus, in terms of well positioning, perforations might occur as far as away from the gas cap but not so close to the water-oil contact to avoid significant water production by coning. The recovery factor from a gas-cap normally will be 20 to 40% – it can be up to 60% in some cases- and the mechanism is dependent upon the following factors:

Size of the gas cap: the degree of pressure maintenance depends upon the gas volume in the gas cap compared to the oil volume.

Vertical permeability: a good vertical permeability will permit the oil moving downward with less bypassing of the gas.

Oil viscosity: an increase of oil viscosity will lead to lower recovery factors, due to the increase in the amount of gas bypassing.

Degree of gas conservation: it is necessary to shut down wells which are producing excessive gas.

Oil Production Rate: lower producing rates will maximize the amount of free gas in the oil zone to migrate to the gas cap which will increase recovery.

Dip angle: steep angle of dip allows better oil drainage.

The recovery factor is higher than for depletion drive mechanisms, given that no gas saturation is formed throughout the reservoir at the same time. Additionally, gas cap reservoirs produce very little or no water.

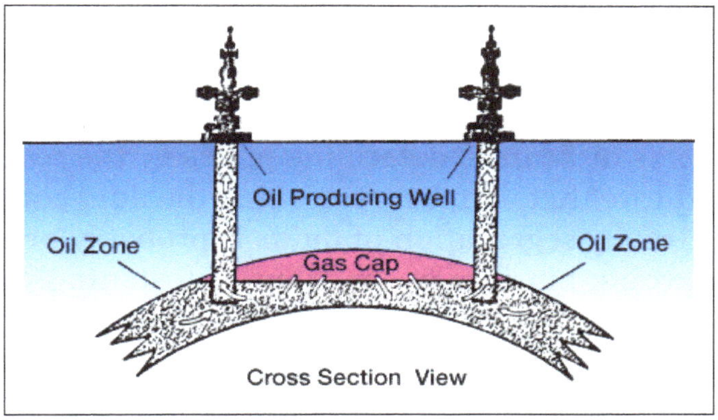

Water Drive

Water drive is a reservoir-drive mechanism whereby the oil is driven through the reservoir by an active aquifer. As the reservoir depletes, the water moving in from the aquifer below displaces the oil until the aquifer energy is expended or the well eventually produces too much water to be viable.

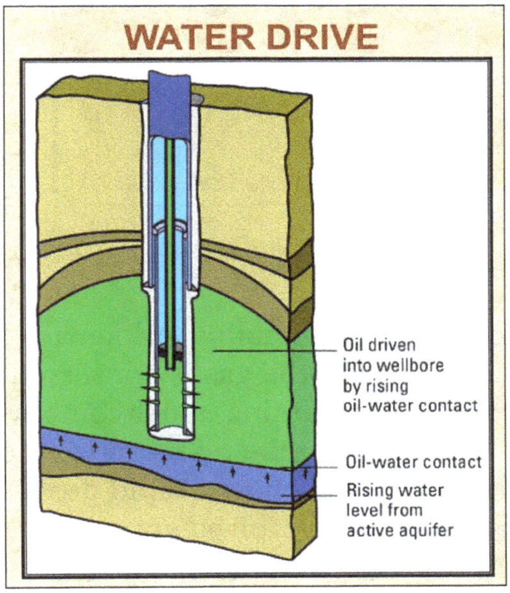

Gravity Drainage

The mechanism of gravity drainage occurs in petroleum reservoirs as a result of differences in densities of the reservoir fluids. The effects of gravitational forces can be simply illustrated by placing a quantity of crude oil and a quantity of water in a jar and agitating the contents. After agitation, the jar is placed at rest, and the denser fluid (normally water) will settle to the bottom of the jar, while the less dense fluid (normally oil) will rest on top of the denser fluid. The fluids have separated as a result of the gravitational forces acting on them. The fluids in petroleum reservoirs have all been subjected to the forces of gravity, as evidenced by the relative positions of the fluids, i.e., gas on top, oil underlying the gas, and water underlying oil. Due to the long periods of time involved in the petroleum accumulation-and-migration process, it is generally assumed that the reservoir fluids are in equilibrium. If the reservoir fluids are in equilibrium, then the gas-oil and oil water contacts should be essentially horizontal. Although it is difficult to determine precisely the reservoir fluid contacts, best available data indicate that, in most reservoirs, the fluid contacts actually are essentially horizontal. Gravity segregation of fluids is probably present to some degree in all petroleum reservoirs, but it may contribute substantially to oil production in some reservoirs.

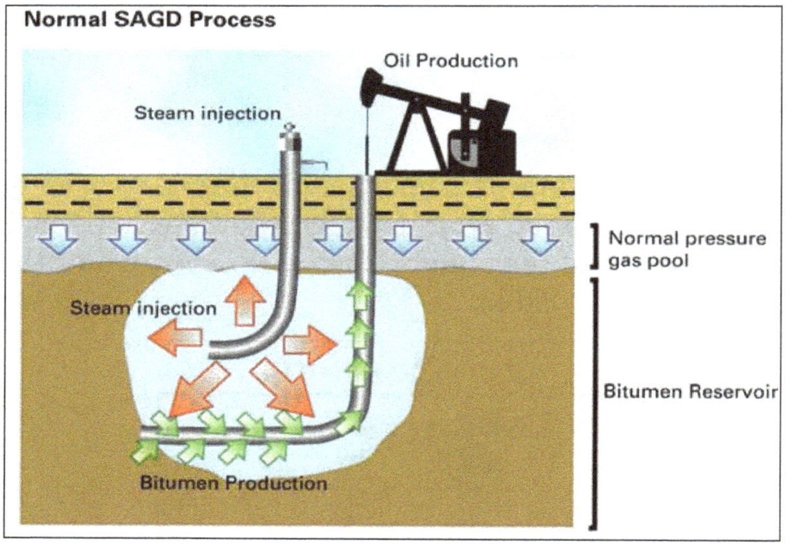

Reservoir Pressure

Variable rates of pressure decline, depending principally upon the amount of gas conservation. Strictly speaking, where the gas is conserved and reservoir pressure is maintained, the reservoir would be operating under combined gas-cap drive and gravity-drainage mechanisms. Therefore, for the reservoir to be operating solely as a result of gravity drainage, the reservoir would show a rapid pressure decline. This would require the up structure migration of the evolved gas where it later would be produced from structurally high wells, resulting in rapid loss of pressure.

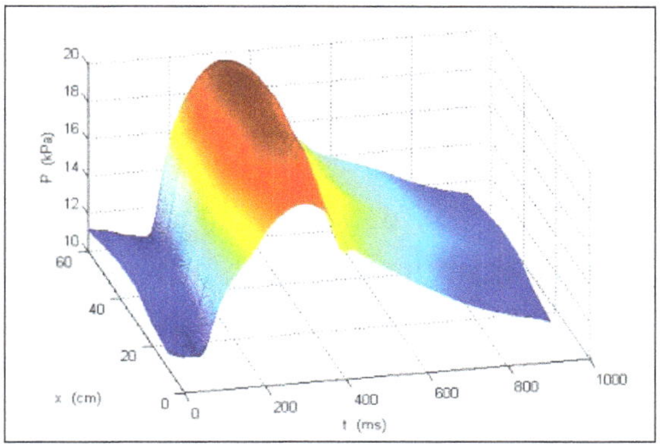

Gas-Oil Ratio

Low gas-oil ratio from structurally low wells. This is caused by migration of the evolved gas up structure due to gravitational segregation of the fluids. On the other hand, the structurally high wells will experience an increasing gas-oil ratio as a result of the up structure migration of the gas released from the crude oil.

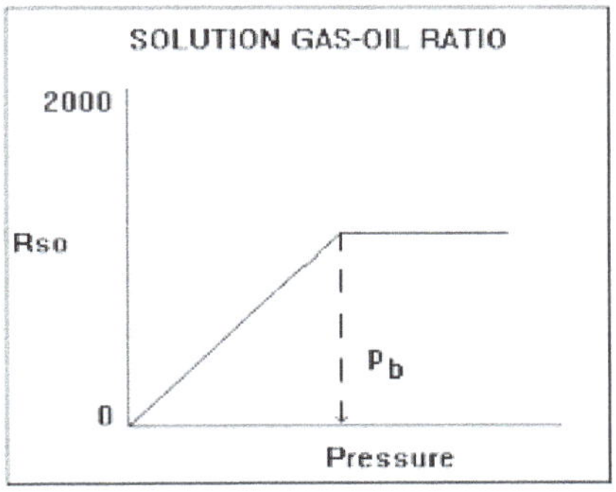

Secondary Gas Cap

Formation of a secondary gas cap in reservoirs that initially were under saturated. Obviously the gravity-drainage mechanism does not become operative until reservoir pressure has declined below the saturation pressure, since above the saturation pressure there will be no free gas in the reservoir.

Formation of a Secondary Gas Cap during gas solution liberation

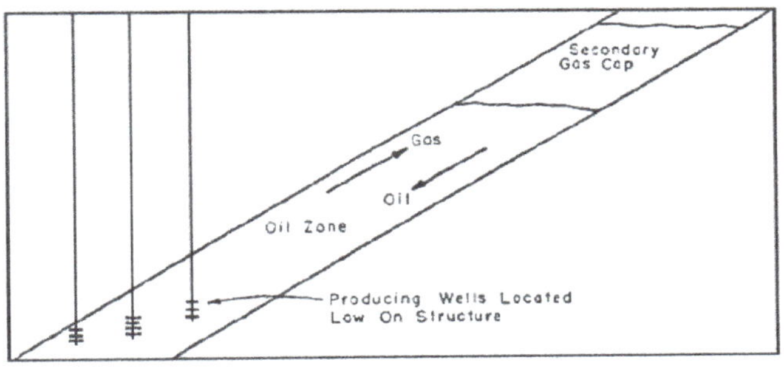

Water Production

Little or no water production. Water production is indicative of a water drive.

Ultimate Oil Recovery

Ultimate recovery from gravity-drainage reservoirs will vary widely, due primarily to the extent of depletion by gravity drainage alone. Where gravity drainage is good, or where producing rates are restricted to take maximum advantage of the gravitational forces, recovery will be high. There are reported cases where recovery from gravity-drainage reservoirs has exceeded 80% of the initial oil in place. In other reservoirs where depletion drive also plays an important role in the oil recovery process, the ultimate recovery will be less.

In operating a gravity-drainage reservoir, it is essential that the oil saturation in the vicinity of the well bore must be maintained as high as possible. There are two basic reasons for this requirement:

- A high oil saturation means a higher oil flow rate.
- A high oil saturation means a lower gas flow rate.

If the evolved gas migrates up structure instead of toward the well bore, then high oil saturation in the vicinity of the well bore can be maintained. In order to take maximum advantage of the gravity-drainage-producing mechanism, wells should be located as structurally low as possible.

Factors that affect ultimate recovery from gravity-drainage reservoirs are:

- Permeability in the direction of dip.
- Dip of the reservoir.
- Reservoir producing rates.
- Oil viscosity.
- Relative permeability characteristics.

Combination or Mixed Drive

The driving mechanism most commonly encountered is one in which both water and free gas are available in some degree to displace the oil toward the producing wells. The most common type of drive encountered, therefore, is a combination-drive mechanism as illustrated in Figure. Two combinations of driving forces can be present in combination drive reservoirs. These are:

(1) Depletion drive and a weak water drive and;

(2) Depletion drives with a small gas cap and a weak water drive.

Then, of course, gravity segregation can play an important role in any of the aforementioned drives.

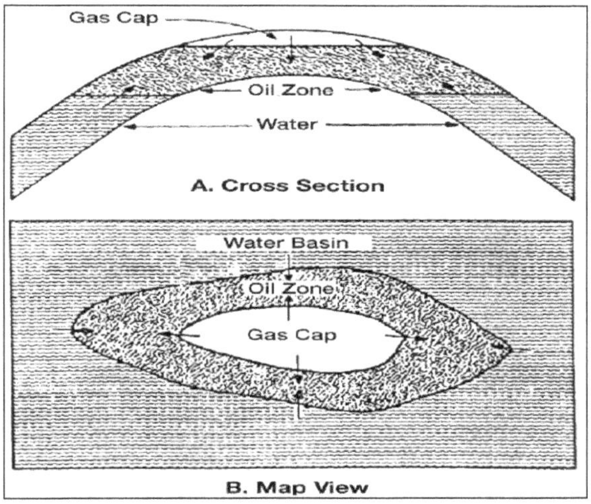

Combination-drive reservoirs can be recognized by the occurrence of a combination of some of the following factors:

a. Relatively rapid pressure decline. Water encroachment and external gas-cap expansion are insufficient to maintain reservoir pressures.

b. Water encroaching slowly into the lower part of the reservoir. Structurally low producing wells will exhibit slowly increasing water producing rates.

c. If a small gas cap is present the structurally high wells will exhibit continually increasing gas-oil ratios, provided the gas cap is expanding. It is possible that the gas cap will shrink due to production of excess free gas, in which case the structurally high wells will exhibit a decreasing gas-oil ratio. This condition should be avoided whenever possible, as large volumes of oil can be lost as a result of a shrinking gas cap.

d. A substantial percentage of the total oil recovery may be due to the depletion-drive mechanism. The gas-oil ratio of structurally low wells will also continue to increase due to evolution of solution gas throughout the reservoir, as pressure is reduced.

e. Ultimate recovery from combination-drive reservoirs is usually greater than recovery from depletion-drive reservoirs but less than recovery from water-drive or gas-cap-drive reservoirs. Actual recovery will depend upon the degree to which it is possible to reduce the magnitude of recovery by depletion drive. In most combination-drive reservoirs, it will be economically feasible to institute some type of pressure maintenance operation, either gas injection, water injection, or both gas and water injection, depending upon the availability of the fluids.

Secondary Recovery

Secondary oil recovery is employed when the pressure inside the well drops to levels that make primary recovery no longer viable. Pressure is the key to collecting oil from the natural underground rock formations in which it forms. When a well is drilled, the pressure inside the formation pushes the oil deposits from the fissures and pores where it collects and into the well where it can be recovered.

But this initial pressure is finite. In order to continue collecting the oil, the pressure must be maintained through other means. These tactics are referred to as secondary recovery techniques.

Methods of Secondary Recovery

There are three main methods of secondary recovery: thermal recovery, gas injection and chemical injection.

The most widely used method of secondary oil recovery is gas injection. Once gas, such as nitrogen or carbon dioxide, is introduced into the reservoir, it expands. This expansion forces oil through the formation and into the well. Gas injection accounts for 60 percent of secondary oil recovery in the U.S.

Thermal recovery is used in about 40 percent of oil wells and relies on heat to facilitate production. Injecting steam or heated water into the reservoir lowers the viscosity of oil, thinning it so that it flows more easily through rock formations and into the well.

The third method, chemical injection, is the least prevalent and is used in less than 1 percent of U.S. oil wells. This approach, also known as chemical flooding, uses solutions composed of micellar polymers and water to reduce the friction between oil and water. Like thermal recovery, this lowers the viscosity of the oil and increases flow.

Secondary recovery techniques are increasingly important in oil and gas exploration since employing these methods can result in a well-being up to 40 percent more productive than with primary methods alone.

Waterflooding

Waterflooding is also known as Water injection. It is a form of this secondary EOR production process.

Used in onshore and offshore developments, water injection involves drilling injection wells into a reservoir and introducing water into that reservoir to encourage oil production. While the injected water helps to increase depleted pressure within the reservoir, it also helps to move the oil in place.

Whether water injection occurs after production has already been depleted or before production from the reservoir has been drained, waterflood sweeps remaining oil through the reservoir to production wells, where it can be recovered.

Water Injection Methods

The water used for water injection is usually some sort of brine, but it can also be made up of other sources that are treated. For example, in some reservoirs water is produced with the hydrocarbons, removed from the production and re-injected into the formation.

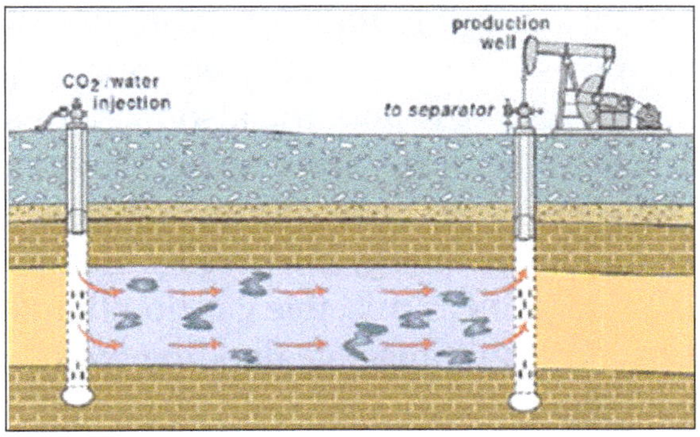

It is important that the water being injected works within the formation. Filtration and processing of the water that will be injected are sometimes necessary to ensure that no materials clog the well pores and that bacteria is not permitted to grow. In an effort to reduce any corrosion within the reservoir, oxygen is often removed from the water, as well.

While production wells can be converted into injection wells, water-injection wells are also drilled specifically for this purpose. Water is then pumped into the reservoir, or gravity can help to push the liquid into the formation. This solution positions water tanks on hills or somewhere above the well, and the water simply is fed into the wellbore.

There are a number of techniques for determining where the water-injection wells should be drilled, as well as established patterns for water-injection wells in relation to production wells. One popular pattern, called the five-spot pattern, involves drilling four water-injection wells in a square around a production well. This is repeated around each production well on the reservoir, resulting in four production wells surrounding each water-injection well, as well.

Other drilling techniques include the seven-spot pattern, which has six water-injection wells surrounding a production well, and the inverted seven-spot pattern, which describes six production wells surrounding one water-injection well.

Also, wells can be drilled in line patterns, rather than spot patterns, where a direct line or staggered line of production wells is followed by a similar line of water-injection wells, and so on. In an edge waterflood, water-injection wells are drilled along the outside borders of the field, and water is injected, with production flowing toward the production wells in the center of the reservoir.

Gasflooding

Gas flooding is also known as miscible flooding. It is one of the leading enhanced oil recovery (EOR) technologies employed for recovering oil that was formerly referred to as either stranded or trapped. Gas flooding is an "enhanced oil recovery" application for injecting miscible (and immiscible) gases into an oil reservoir for increasing oil production.

Gas flooding typically includes CO_2, natural gas or nitrogen as the gas that is injected. Gas flooding tasked place as either a miscible flood or an immiscible flood. Miscible means that the gas that is injected "mixes" with the oil, thereby reducing viscosity and interfacial tension of the oil and rock. Miscible gas flooding also increases oil "swelling" and localized pressure or drive within the reservoir. "Immiscible" flooding means that the gas that is injected into the reservoir does not mix or go into solution. Therefore the purpose of the immiscible flood is to provide the energy or drive by increased pressure. Immiscible flooding does not produce as much oil as miscible gas flooding; however there are certain applications and reservoirs wherein immiscible flooding is well-suited.

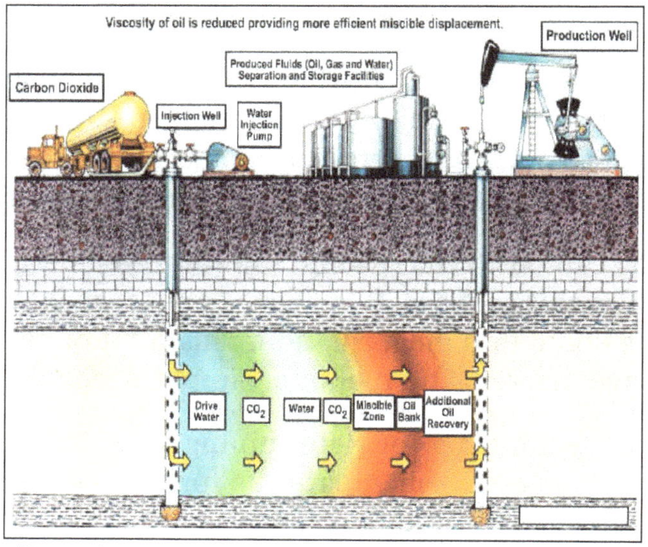

Gas Flooding using CO_2 (CO_2-EOR) as the "miscible" gas for injection.

Tertiary Recovery

Tertiary recovery is the third phase of oil extraction from an oil reserve. This phase of removal allows petroleum companies to remove a significant amount of oil from a reserve which they would not be able to access without these enhanced methods.

There are three primary methods of tertiary recovery:

1. With thermal recovery, the reservoir is heated, often with the introduction of steam. This warms the oil, thinning it so that it loses some of its viscosity and is more apt to flow.
2. In gas injection, the pumping of gases, such as carbon dioxide, nitrogen, or natural gas, into the reservoir is used. The gases expand, and the pressure pushes the remaining oil through the reservoir.
3. Using chemical injection involves injecting polymers, which are long-chained molecules, into the reservoir to lower surface tension and allow the oil to flow more freely. This method is used significantly less frequently than thermal recovery or gas injection.

Thermal EOR

Thermal EOR methods are i) cyclic steam injection (sometimes referred to as Huff & Puff), ii) steamflooding, and iii) Steam- Assisted Gravity Drainage (SAGD).

Cyclic Steam Injection

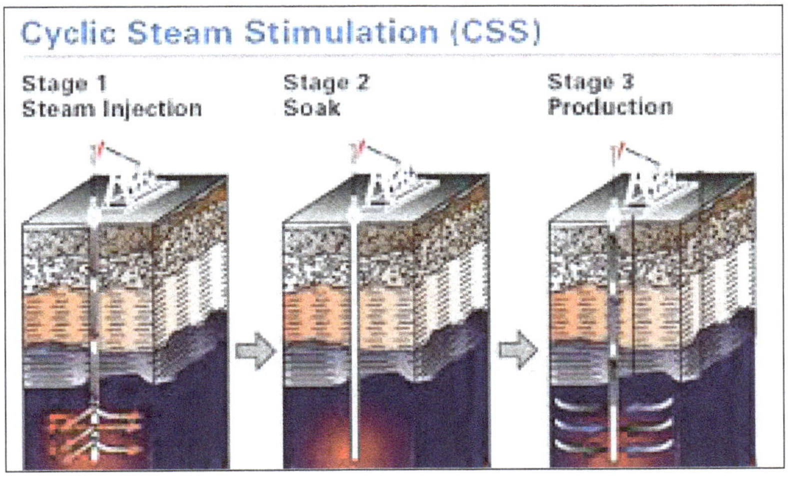

Cyclic steam injection (CSI) is usually done in 3 stages: injection, soaking, and production. First, a predetermined amount of steam is injected into the well to heat the oil in the surrounding reservoir (injection stage). Once the desired amount of steam is injected, the well is shut down to allow the steam to heat reservoir around the well (soaking stage). In the last stage, the injection well is converted to a production mode until the heat is dissipated with the produced fluids (production stage). This cycle is repeated until the response becomes insignificant and economical limits are reached. Obviously, most of the oil is produced in the first few cycles.

CSI is most commonly applied to the reservoirs with thickness greater than 30 ft and depth less than 3000 foot; high reservoir porosity and oil saturation are desirable. Typical recovery factors for this method are in the range of 10-30%. It is very common for wells to produce for a few Huff&Puff cycles before switching to a different thermal EOR method, namely steamflooding.

Steamflooding

For steamflood operations, some wells are used for injection and others are used for production. During the steamflood, high-quality steam is injected into the heavy-oil reservoir; steam heats the oil and pushes it toward a producing well. As opposed to the CSI, during the steamflood, two mechanisms are involved in the recovery process. As in the case of any other thermal method, the first mechanism is viscosity reduction due to increase of temperature; the second mechanism is a physical displacement of oil by steam and hot water. The recovery factors for the steamflood operations are usually higher than one for the CSI and are in the range of 40-60%.

As steamflood matures, a large amount of heat is retained in the reservoir rock and it may become uneconomical to inject steam into the reservoir. In this case the steamflood project is usually stopped or field is converted into a waterflood mode.

Steam-Assisted Gravity Drainage

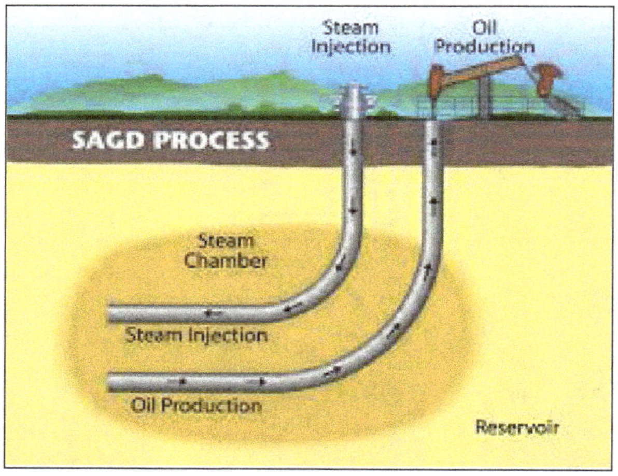

Steam-assisted gravity drainage (SAGD) was first proposed and developed by R. Butler and his coworkers in Imperial Oil. The Foster Creek plant in Alberta Canada was the first commercial SAGD project. Since Foster Creek plant, SAGD has been used increasingly for the recovery of bitumen and heavy oil in Canada. SAGD is considered to be an advanced form of steam injection: here, two horizontal wells are placed, one a few meters above another. The upper well is used as a steam injector; whereas, another well is used as a producer. The steam from the injection well flows towards the perimeter of a steam chamber and condenses. The steam chamber grows both vertically and horizontally heating the surrounding oil with subsequent reduction of oil viscosity. Heated oil and steam condensate flow toward the production well due to gravity.

Method	Location	SOR
Steam Floods	California, USA	~ 4
CSI	California, USA	1.0 - 2.0
CSI	Alberta, Canada	2.0 - 3.3
CSI	Venezuela	~ 0.33
SAGD	Alberta, Canada	2.0 - 3.3

Generally, SAGD leads to a higher recovery factors (about 40-60%) compared to the CSI. These high recoveries are obtained because of the systematic nature of the drainage process. Several different reservoir cutoffs for economical SAGD operations are described by McCorkack. Among the most restrictive cutoffs are: pay thickness greater than 40 ft, permeability greater than 3 Darcy, and absence of top/bottom water. Currently, there is an attempt to improve SAGD technology by introducing noncondensible gas to the steam stream as in SAGP process; or by injecting solvent vapor along with noncondensible gas as in Vapex process. A common feature of the three processes (SAGD, SAGP, and VAPEX) is the use of a pair of long horizontal wells.

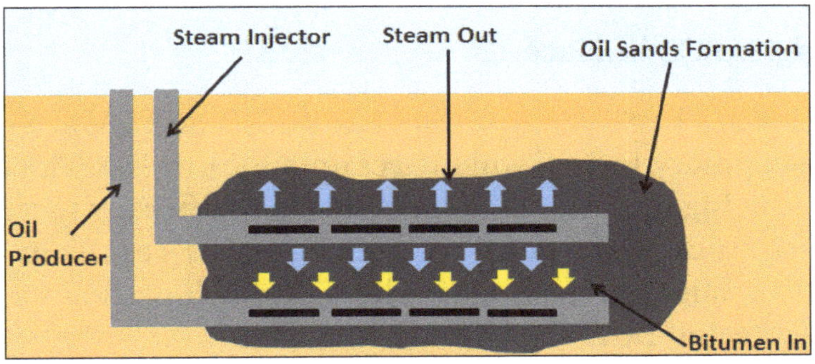

Energy Efficiency

All of the methods described above require steam to be generated on the surface and injected into the petroleum reservoir. Most of the energy consumed during thermal EOR is due to steam generation. The amount of oil produced per amount of steam injected determines the method energy efficiency. For the steam based methods, the commonly used metrics are the volumetric steam-to-oil ratio (SOR) and oil-to-steam ratio (OSR). Since the amount of oil produced and the amount of steam injected vary in time, sometimes it is more meaningful to use cumulative volumes instead of instantaneous. The SOR's are strongly dependent on the quality (density, viscosity, etc) of the produced hydrocarbons, reservoir properties (porosity, permeability), and surface facilities being used. The steam-to-oil ratios for CSI, steamflooding, and SAGD at different locations are presented in Table above.

One of the ways to reduce SOR for a given reservoir is to optimize the design of the surface facilities. For instance, it is beneficial to minimize transportation of hot water to avoid heat losses; or to maximize heat integration between hot and cold process streams to minimize external cooling. On the reservoir side, several optimizations can be done: well placement optimization, injection strategy/production strategy optimization, etc. Commonly, both reservoir and surface facilities optimization are performed during the conceptual stage of the project.

Chemical EOR

Chemical enhanced oil recovery is an advanced technology that addresses the mechanisms that produce oil. The standard way to view oil production is though the oil recovery equation shown below.

$$Np = ED\ EV \bullet OOIP$$

Where:

- Np – Oil Recovery (Production).
- ED – Pore to Pore (Unit) Displacement Efficiency.
- EV – Volumetric Sweep Efficiency.
- OOIP – Original Oil in Place.

Typical oil recoveries are between 20-45% because ED and EV are low. Oil is trapped in the pores due to capillary forces, which accounts for low ED. The capillary force can be minimized by reducing the interfacial tension (IFT), thereby increasing ED. At high IFT the oil drops are stuck in the pores; reducing the IFT results in a more flexible oil drop that is mobile through the reservoir. The figure below shows a simplistic view of how the oil drop is affected by IFT.

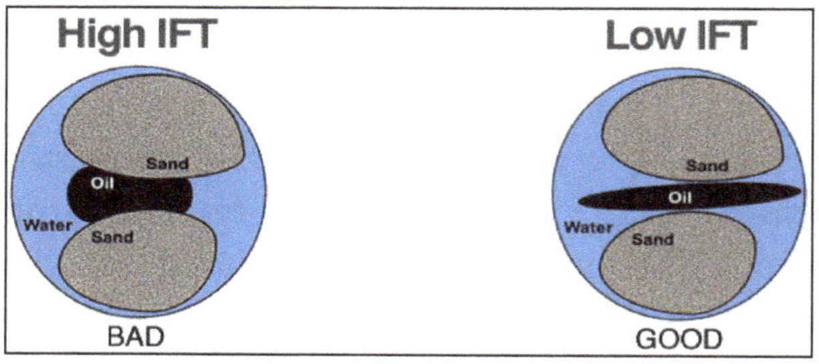

EV is based on water's ability to sweep oil to a producer. The key concept with EV is mobility ratio (M). A simplistic way to understand M is to view it as the ratio of the velocity of water to the velocity of oil. If water moves faster than the oil (high M), it will bypass the oil and break through faster than desired (fingering and channeling affects). If the water moves at the same speed as the oil, it will do a better job sweeping the oil. The figure below shows the effect M has in displacing oil. Polymer agents are added to injection water to viscosify the water and lower its mobility to a more favorable condition.

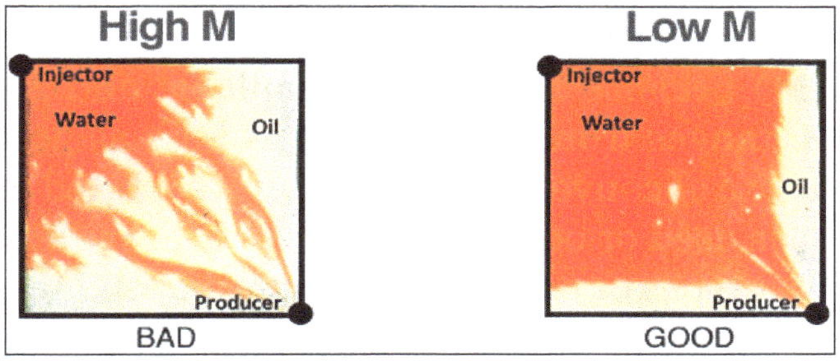

Chemical EOR represents the best methods of improving both efficiencies, ED and EV.

CEOR uses Surfactants and Alkali, and Polymers in an aqueous solution to improve recovery efficiency. Alkalis and Surfactants improve ED, mobilizing the oil that is trapped in the pores. Polymers improve EV by enhancing the water's ability to sweep oil to neighboring production wells by increasing its viscosity.

Each field has a unique oil, water, rock and temperature environment, and therefore the CEOR formulation is specifically developed.

Unlike other forms of EOR, such as CO2 which requires a nearby source, chemical EOR can be applied anywhere chemical can be transported by road, rail, or boat.

Chemical EOR is implemented in stages and actual size/duration of each stage depends on economics.

Typical chemical flood incremental recovery is in the range of 5-30% OOIP, depending on the process selected and the characteristics of the reservoir.

The figure below shows the stage of implementation for an ASP flood.

1. ASP stage – mobilizes oil trapped by capillary forces creating an oil bank.
2. Polymer stage – pushes oil bank to producer.
3. Water stage – final push to finish the project.

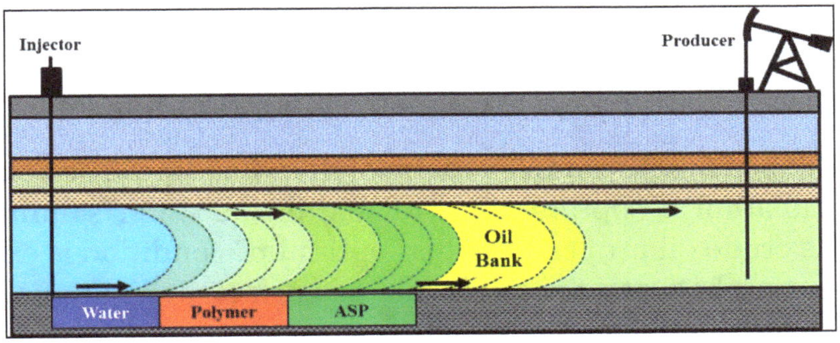

Mobility Control Polymer

Polymer Flooding (P) is the addition of water soluble polymer to viscosify injection water in order to reduce its mobility relative to the oil. Reduction of the mobility attempts to overcome poor sweep efficiency that occurs when viscous oil is displaced with lower viscosity water. Most polymer floods use partially hydrolyzed polyacrylamide polymer (HPAM) because it is widely available and relatively inexpensive. Polymer concentration ranges from 500 ppm to as high as 3000 ppm in reservoirs with highly viscous oil. Use of conventional polymers is limited to ~80°C unless water with very low hardness is used or other chemicals agents are included to stabilize the polymer. However, other types of polymers can push beyond temperature and chemical limitations of HPAM. These include AMPS (copolymer with sulfuric acid) and NVP (with N-vinyl pyrrolidone). Permeability is limited to a minimum of ~25 mD because polymer molecules are large enough to plug the small pore throats found in low permeability formations.

Polymer flooding has been applied successfully throughout the world for over 40 years and it is currently the most widely used chemical flooding technology based on the number of field applications and the mass of polymer injected.

Surfactant Polymer

Surfactant-Polymer (SP) involves adding low concentrations (0.1% to 2%) of a surfactant and co-surfactant and co-solvent and salt to the injection water in order to reduce the oil-water interfacial tension. Polymer is added to increase the solution viscosity in order to overcome viscous instability of low interfacial tension displacement. The SP solution is injected for 20% to 40% pore volume of the target oil-bearing zone, followed by similar volumes of polymer flush.

Conceptually, SP flooding addresses both sweep efficiency and displacement efficiency, limited only to those conditions limiting polymer flooding. Practically, the process is limited to the amount of surfactant and polymer that can be blended together because at some concentration of one in the presence of the other, the solution begins to separate. The concentration of surfactant used must be greater than the critical micelle concentration (CMC) at which point the interfacial tensions become very low for surfactant at "optimum" conditions. The CMC is often quite low. However, both polymer and, to a much greater degree, surfactant are adsorbed by the reservoir rock surface, which decreases their concentration as they progress through the reservoir. Thus the surfactant concentration must be high enough to overcome this adsorption loss.

Designing an SP flood is more complex than polymer flooding. The condition at which a surfactant formulation is "optimum" is related to the water salinity, temperature, and pH, as well as crude oil properties. This "optimum" condition is expressed by very low interfacial tension between the crude oil and surfactant solution. Certain types of surfactant can handle the different reservoir temperature, water salinity, etc. or other unfavorable conditions, but no surfactant is the silver bullet that can handle them all.

Micellar-Polymer

Micellar-Polymer (MP) is similar to SP except higher concentrations of surfactant are used (2% to 12%) and are injected for lower pore volume (5% to 20%). This structuring and many other properties of micellar solutions are sensitive to changes in salinity, temperature, etc. The micellar "slug" is followed by a polymer flush as with SP process.

Micellar-Polymer flooding was widely researched by major oil companies as a means to overcome the limitations of surfactant-polymer flooding – principally the adsorption loss of surfactant. The process proved to be technically very successful in field scale demonstration projects by many major oil companies in the 70's and 80's (10% OOIP to 20% OOIP incremental oil), but was then, and would be now uneconomic due to the high concentrations of surfactant used and their associated costs.

Alkali-Polymer

Alkali-Polymer (AP) flooding involves adding an alkaline agent along with polymer to softened injection water. The water must be softened because the alkaline agents would cause any divalent cations, such as calcium and magnesium, to precipitate, and this solid precipitate will plug most formations. The alkali reacts with components of some oils (saponify) to form "soap" that in the right environment will reduce the interfacial tension sufficiently to overcome capillary forces trapping the oil. The AP solution is injected for 20% to 40% pore volume and followed by a similar volume of polymer flush.

AP flooding has been and is currently being used in full field projects with incremental recoveries of 10% OOIP to 20% OOIP. Of all the chemical flooding processes, it probably has the most limited number of potential applications because not all oils have components that saponify. Also, softening water itself can be complex and costly, depending upon the water hardness and salinity. However, AP flooding is more economic than processes using surfactant because the cost of alkali is much lower than surfactant.

Alkali-Surfactant-Polymer

Alkali-Surfactant-Polymer (ASP) flooding simply adds surfactant (0.05% to 0.5%) to an AP solution to broaden the range of reservoir environment for which the ASP process applies. Many light oils do not contain sufficient amounts of the components that react with alkali to reduce the oil-water interfacial tension sufficiently to overcome capillary forces trapping the oil. Blending surfactant with the alkali overcomes this barrier. As with Alkali-Polymer flooding, the water must have very low divalent ion content (hardness), often requiring softening.

One property of alkaline agents is that they change the reservoir rock chemistry in such a way as to significantly decrease polymer and surfactant adsorption. Therefore, ASP projects use lower concentrations of added surfactant than Surfactant-Polymer or Micellar-Polymer projects. ASP flooding is being applied in field projects around the world with incremental recoveries in some cases greater than 20% OOIP.

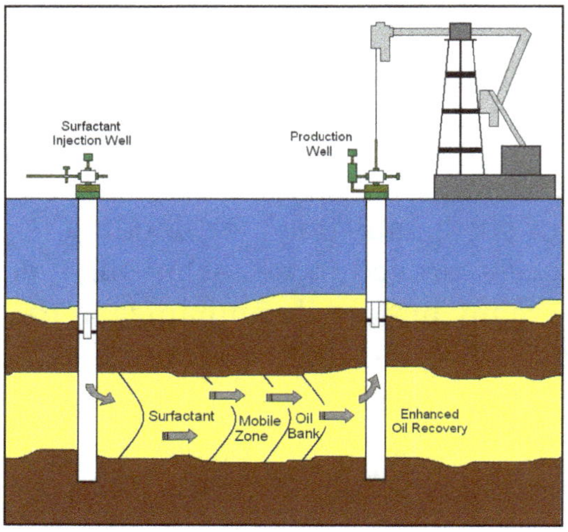

Miscible Gas Flooding

Miscible gas flooding method uses a fluid that is miscible with the oil. Such a fluid has a zero interfacial tension with the oil and can in principal flush out all of the oil remaining in place. In practice a gas is used since gases have high mobilities and can easily enter all the pores in the rock providing the gas is miscible in the oil. Three types of gas are commonly used:

(i) CO_2

(ii) N_2

(iii) Hydrocarbon gases

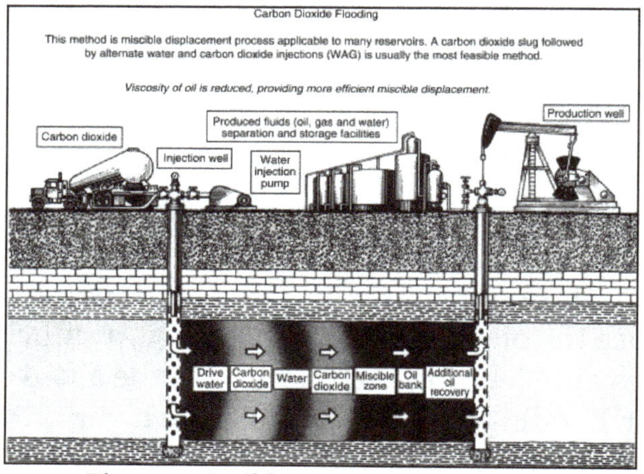

Figure: Miscible WAG Flooding EOR.

All of these are relatively cheap to obtain either from the atmosphere or from evolved reservoir gases. The high mobility of gases can cause a problem in the reservoir flooding process, since gas breakthrough may be early due to fingering, leading to low sweep efficiencies. Effort is then concentrated on trying to improve the sweep efficiency. One such approach is called a miscible WAG (water alternating gas). In this approach water slugs and CO_2 slugs are alternately injected into the reservoir; the idea being that the water slugs will lower the mobility of the CO_2 and lead to a more piston-like displacement with higher flood efficiencies. An additional important advantage of miscible gas-flooding is that the gas dissolves in the oil, and this process reduces the oil viscosity, giving it higher mobilities and easier recovery.

Infill Recovery

Infill drilling is a practice used to speed-up the rate of hydrocarbon recovery by finding the best locations to insert new producers and injectors within the original well pattern. Due to reservoir heterogeneity, wells drilled into a field may not be doing enough to drain hydrocarbons. This can lead to inefficient recovery and sub-optimal reservoir performance.

Infill drilling looks for "sweet spots" within the reservoir that has been by-passed or poorly drained by existing wells, then drills new wells into these spots. This practice is particularly important to reservoirs with low permeability as adding new wells within the original well pattern can help speed up recovery and even add new reserves. In essence, infill drilling reduces the distance between wells for efficient reservoir sweep.

Before Drilling Infill Wells

A successful infill well will boost hydrocarbon production.

It is important to understand that infill drilling is applied in mature fields. It is done after the wells in the field have already been producing for a while and a need arises to add more wells. The practice of infill drilling begins with data gathering and data interpretation for proper reservoir description.

This step is crucial in obtaining an accurate understanding of reservoir connectivity and flow barriers. 4D seismic, which gives a graphic view of where reservoir fluids (oil, gas and water) have accumulated with time, can be used to pin-point the right location to drill infill wells.

Infill drilling is not limited to only production wells; infill wells can also be injection wells drilled to sweep by-passed oil zones towards producing wells.

Locating Infill Wells

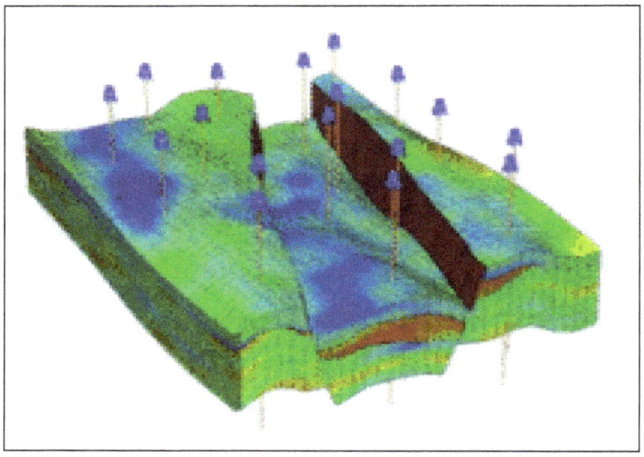

Data gathering through 4D seismic can help in locating sweet spots and by-passed zones.

The choice of "where to drill" an infill well is not as simple as it seems. Multiple possible realistic models should be constructed to determine the optimum location for a new well. This involves testing hundreds or even thousands of potential infill alternatives to know which one will perform best.

Wrongly locating infill wells will result in the new well draining the same area as a previous well thereby eliminating any benefit from the infill well. A poorly located infill injection well may not even be communicating with zones containing hydrocarbons at all leading to a channeling of the injected fluid to wrong zones.

Drilling too many infill wells than necessary is not economical and drilling too few infill wells will result in poor hydrocarbon recovery. Optimum well spacing and location is the goal.

References

- Hydrocarbon-recovery-mechanisms: petroleumandgasengineering.blogspot.com, Retrieved 11, March 2020

- Drive-mechanisms-in-reservoir: kraken.com.br, Retrieved 26, April 2020
- The-gravity-drainage-drive-mechanism: assignmenthelp.net, Retrieved 30,June 2020
- Primary-recovery: investopedia.com, Retrieved 26,May 2020
- Solution-gas-drive-mechanism: drillingformulas.com, Retrieved 13,July 2020

6

Petroleum Refinery and its Processes

The industrial process plant which transforms crude oil into various useful products like petroleum, gasoline, kerosene, liquefied petroleum gas, etc. This chapter covers amine gas treating, catalytic reforming, cracking, fluid catalytic cracking, gas flare, hydrodenitrogenation and hydrodesulfurization in an easy-to-understand manner. The topics elaborated in this chapter will help in gaining a better perspective about petroleum refinery.

Oil Refinery

An oil refinery or petroleum refinery is an industrial processplant where crude oil is processed and refined into more useful products such as petroleum naphtha, gasoline, diesel fuel, asphalt base, heating oil, kerosene, and liquefied petroleum gas. Oil refineries are typically large, sprawling industrial complexes with extensive piping running throughout, carrying streams of fluids between large chemical processing units. In many ways, oil refineries use much of the technology of, and can be thought of, as types of chemical plants. The crude oil feed stock has typically been processed by an oil production plant. There is usually an oil depot (tank farm) at or near an oil refinery for the storage of incoming crude oil feedstock as well as bulk liquid products.

Anacortes Refinery (Tesoro), on the north end of March Point southeast of Anacortes, Washington.

An oil refinery is considered an essential part of the downstream side of the petroleum industry.

Operation

Raw or unprocessed crude oil is not generally useful in industrial applications, although "light, sweet" (low viscosity, low sulfur) crude oil has been used directly as a burner fuel to produce steam for the propulsion of seagoing vessels. The lighter elements, however, form explosive vapors in the fuel tanks and are therefore hazardous, especially in warships. Instead, the hundreds of different hydrocarbon molecules in crude oil are separated in a refinery into components which can be used as fuels, lubricants, and as feedstocks in petrochemical processes that manufacture such products as plastics, detergents, solvents, elastomers and fibers such as nylon and polyesters.

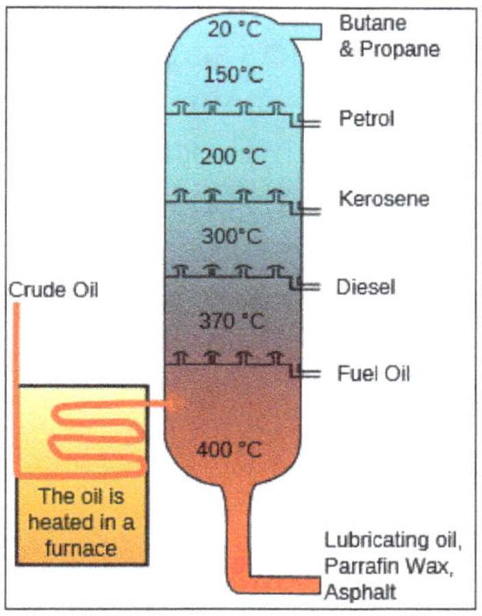

Crude oil is separated into fractions by fractional distillation. The fractions at the top of the fractionating column have lower boiling points than the fractions at the bottom. The heavy bottom fractions are often cracked into lighter, more useful products. All of the fractions are processed further in other refining units.

Petroleumfossil fuels are burned in internal combustion engines to provide power for ships, automobiles, aircraft engines, lawn mowers, dirt bikes, and other machines. Different boiling points allow the hydrocarbons to be separated by distillation. Since the lighter liquid products are in great demand for use in internal combustion engines, a modern refinery will convert heavy hydrocarbons and lighter gaseous elements into these higher value products.

The oil refinery in Haifa, Israel is capable of processing about 9 million tons (66 million barrels) of crude oil a year. Its two cooling towers are landmarks of the city's skyline.

Oil can be used in a variety of ways because it contains hydrocarbons of varying molecular masses, forms and lengths such as paraffins, aromatics, naphthenes (or cycloalkanes), alkenes, dienes, and alkynes. While the molecules in crude oil include different atoms such as sulfur and nitrogen, the hydrocarbons are the most common form of molecules, which are molecules of varying lengths and complexity made of hydrogen and carbonatoms, and a small number of oxygen atoms. The differences in the structure of these molecules account for their varying physical and chemical properties, and it is this variety that makes crude oil useful in a broad range of several applications.

Once separated and purified of any contaminants and impurities, the fuel or lubricant can be sold without further processing. Smaller molecules such as isobutane and propylene or butylenes can be recombined to meet specific octane requirements by processes such as alkylation, or more commonly, dimerization. The octane grade of gasoline can also be improved by catalytic reforming, which involves removing hydrogen from hydrocarbons producing compounds with higher octane ratings such as aromatics. Intermediate products such as gasoils can even be reprocessed to break a heavy, long-chained oil into a lighter short-chained one, by various forms of cracking such as fluid catalytic cracking, thermal cracking, and hydrocracking. The final step in gasoline production is the blending of fuels with different octane ratings, vapor pressures, and other properties to meet product specifications. Another method for reprocessing and upgrading these intermediate products (residual oils) uses a devolatilization process to separate usable oil from the waste asphaltene material.

Oil refineries are large scale plants, processing about a hundred thousand to several hundred thousand barrels of crude oil a day. Because of the high capacity, many of the units operate continuously, as opposed to processing in batches, at steady state or nearly steady state for months to years. The high capacity also makes process optimization and advanced process control very desirable.

Major Products

Petroleum products are usually grouped into four categories: light distillates (LPG, gasoline, naphtha), middle distillates (kerosene, jet fuel, diesel), heavy distillates and residuum (heavy fuel oil, lubricating oils, wax, asphalt). This classification is based on the way crude oil is distilled and separated into fractions (called distillates and residuum) as in the above drawing.

- Liquified petroleum gas (LPG)
- Gasoline (also known as petrol)
- Naphtha
- Kerosene and related jet aircraft fuels
- Diesel fuel
- Fuel oils
- Lubricating oils
- Paraffin wax
- Asphalt and tar
- Petroleum coke
- Sulfur
- Olefines
- Heat and electrical energy

Oil refineries also produce various intermediate products such as hydrogen, light hydrocarbons, reformate and pyrolysis gasoline. These are not usually transported but instead are blended or processed further on-site. Chemical plants are thus often adjacent to oil refineries or a number of further chemical processes are integrated into it. For example, light hydrocarbons are steam-cracked in an ethylene plant, and the produced ethylene is polymerized to produce polyethene.

Because technical reasons and environment protection demand a very low sulfur content in all but the most heavy products, it is transformed to hydrogen sulfide via catalytic Hydrodesulfurization and removed from the product stream via Amine gas treating. Using the so-called Claus process, hydrogen sulfide is afterwards transformed to elementary sulfur to be sold to the chemical industry. The rather large heat energy freed by this process is directly used in the other parts of the refinery. Often an electrical power plant is combined into the whole refinery process to take up the excess heat.

Common Process Units Found in a Refinery

Storage tanks and towers at Shell Puget Sound Refinery (Shell Oil Company), Anacortes, Washington.

- Desalter unit washes out salt from the crude oil before it enters the atmospheric distillation unit.
- Atmospheric distillation unit distills crude oil into fractions.
- Vacuum distillation unit further distills residual bottoms after atmospheric distillation.
- Naphtha hydrotreater unit uses hydrogen to desulfurize naphtha from atmospheric distillation. Must hydrotreat the naphtha before sending to a catalytic reformer unit.
- Catalytic reformer unit is used to convert the naphtha-boiling range molecules into higher octane reformate (reformer product). The reformate has higher content of aromatics and cyclic hydrocarbons). An important byproduct of a reformer is hydrogen released during the catalyst reaction. The hydrogen is used either in the hydrotreaters or the hydrocracker.
- Distillate hydrotreater desulfurizes distillates (such as diesel) after atmospheric distillation.
- Fluid Catalytic Cracker (FCC) unit upgrades heavier fractions into lighter, more valuable products.

- Hydrocracker unit uses hydrogen to upgrade heavier fractions into lighter, more valuable products.
- Visbreaking unit upgrades heavy residual oils by thermally cracking them into lighter, more valuable reduced viscosity products.
- Merox unit treats LPG, kerosene or jet fuel by oxidizing mercaptans to organic disulfides.
- Alternative processes for removing mercaptans are known, e.g. doctor sweetening process and caustic washing.
- Coking units (delayed coking, fluid coker, and flexicoker) process very heavy residual oils into gasoline and diesel fuel, leaving petroleum coke as a residual product.
- Alkylation unit uses sulfuric acid or hydrofluoric acid to produce high-octane components for gasoline blending.
- Dimerization unit converts olefins into higher-octane gasoline blending components. For example, butenes can be dimerized into isooctene which may subsequently be hydrogenated to form isooctane. There are also other uses for dimerization. Gasoline produced through dimerization is highly unsaturated and very reactive. It tends spontaneously to form gums. For this reason the effluent from the dimerization need to be blended into the finished gasoline pool immediately or hydrogenated.
- Isomerization unit converts linear molecules to higher-octane branched molecules for blending into gasoline or feed to alkylation units.
- Steam reforming unit produces hydrogen for the hydrotreaters or hydrocracker.
- Liquified gas storage vessels store propane and similar gaseous fuels at pressure sufficient to maintain them in liquid form. These are usually spherical vessels or "bullets" (i.e., horizontal vessels with rounded ends).
- Storage tanks store crude oil and finished products, usually cylindrical, with some sort of vapor emission control and surrounded by an earthen berm to contain spills.
- Amine gas treater, Claus unit, and tail gas treatment convert hydrogen sulfide from hydrodesulfurization into elemental sulfur.
- Utility units such as cooling towers circulate cooling water, boiler plants generates steam, and instrument air systems include pneumatically operated control valves and an electrical substation.
- Wastewater collection and treating systems consist of API separators, dissolved air flotation (DAF) units and further treatment units such as an activated sludge biotreater to make water suitable for reuse or for disposal.
- Solvent refining units use solvent such as cresol or furfural to remove unwanted, mainly aromatics from lubricating oil stock or diesel stock.
- Solvent dewaxing units remove the heavy waxy constituents petrolatum from vacuum distillation products.

Flow Diagram of Typical Refinery

The image below is a schematic flow diagram of a typical oil refinery that depicts the various unit processes and the flow of intermediate product streams that occurs between the inlet crude oil feedstock and the final end products. The diagram depicts only one of the literally hundreds of different oil refinery configurations. The diagram also does not include any of the usual refinery facilities providing utilities such as steam, cooling water, and electric power as well as storage tanks for crude oil feedstock and for intermediate products and end products.

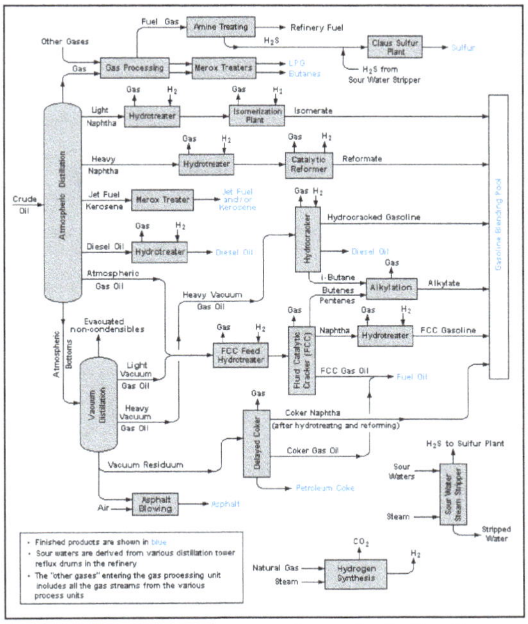

Schematic flow diagram of a typical oil refinery.

There are many process configurations other than that depicted above. For example, the vacuum distillation unit may also produce fractions that can be refined into endproducts such as: spindle oil used in the textile industry, light machinery oil, motor oil, and various waxes.

The Crude Oil Distillation Unit

The crude oil distillation unit (CDU) is the first processing unit in virtually all petroleum refineries. The CDU distills the incoming crude oil into various fractions of different boiling ranges, each of which are then processed further in the other refinery processing units. The CDU is often referred to as the *atmospheric distillation unit* because it operates at slightly above atmospheric pressure.

Below is a schematic flow diagram of a typical crude oil distillation unit. The incoming crude oil is preheated by exchanging heat with some of the hot, distilled fractions and other streams. It is then desalted to remove inorganic salts (primarily sodium chloride).

Following the desalter, the crude oil is further heated by exchanging heat with some of the hot, distilled fractions and other streams. It is then heated in a fuel-fired furnace (fired heater) to a temperature of about 398 °C and routed into the bottom of the distillation unit.

The cooling and condensing of the distillation tower overhead is provided partially by exchanging heat with the incoming crude oil and partially by either an air-cooled or water-cooled condenser. Additional heat is removed from the distillation column by a pumparound system as shown in the diagram below.

As shown in the flow diagram, the overhead distillate fraction from the distillation column is naphtha. The fractions removed from the side of the distillation column at various points between the column top and bottom are called *sidecuts*. Each of the sidecuts (i.e., the kerosene, light gas oil and heavy gas oil) is cooled by exchanging heat with the incoming crude oil. All of the fractions (i.e., the overhead naphtha, the sidecuts and the bottom residue) are sent to intermediate storage tanks before being processed further.

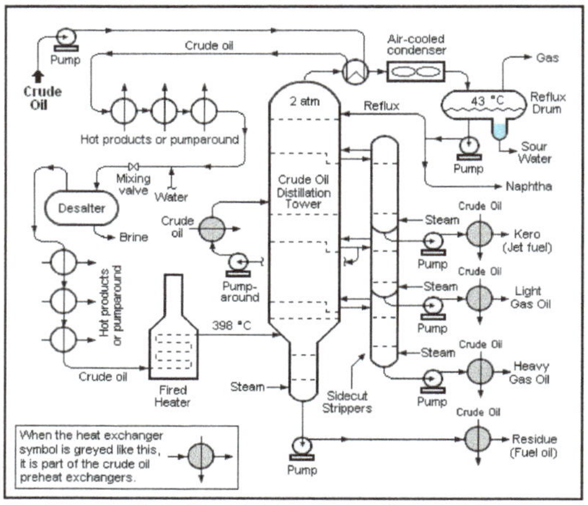

Schematic flow diagram of a typical crude oil distillation unit as used in petroleum crude oil refineries.

Specialty and Products

These require blending various feedstocks, mixing appropriate additives, providing short term storage, and preparation for bulk loading to trucks, barges, product ships, and railcars:

- Gaseous fuels such as propane, stored and shipped in liquid form under pressure in specialized railcars to distributors.
- Lubricants (produces light machine oils, motor oils, and greases, adding viscosity stabilizers as required), usually shipped in bulk to an offsite packaging plant.
- Wax (paraffin), used in the packaging of frozen foods, among others. May be shipped in bulk to a site to prepare as packaged blocks.
- Sulfur (or sulfuric acid), byproducts of sulfur removal from petroleum which may have up to a couple percent sulfur as organic sulfur-containing compounds. Sulfur and sulfuric acid are useful industrial materials. Sulfuric acid is usually prepared and shipped as the acid precursor oleum.
- Bulk tar shipping for offsite unit packaging for use in tar-and-gravel roofing.
- Asphalt unit. Prepares bulk asphalt for shipment.
- Petroleum coke, used in specialty carbon products or as solid fuel.

- Petrochemicals or petrochemical feedstocks, which are often sent to petrochemical plants for further processing in a variety of ways. The petrochemicals may be olefins or their precursors, or various types of aromatic petrochemicals.

Siting/Locating of Petroleum Refineries

A party searching for a site to construct a refinery or a chemical plant needs to consider the following issues:

- The site has to be reasonably far from residential areas.
- Infrastructure should be available for supply of raw materials and shipment of products to markets.
- Energy to operate the plant should be available.
- Facilities should be available for waste disposal.

Refineries which use a large amount of steam and cooling water need to have an abundant source of water. Oil refineries therefore are often located nearby navigable rivers or on a sea shore, nearby a port. Such location also gives access to transportation by river or by sea. The advantages of transporting crude oil by pipeline are evident, and oil companies often transport a large volume of fuel to distribution terminals by pipeline. Pipeline may not be practical for products with small output, and rail cars, road tankers, and barges are used.

Petrochemical plants and solvent manufacturing (fine fractionating) plants need spaces for further processing of a large volume of refinery products for further processing, or to mix chemical additives with a product at source rather than at blending terminals.

Safety and Environmental Concerns

Fire-extinguishing operations after the Texas City refinery explosion.

The refining process releases a number of different chemicals into the atmosphere and a notable odor normally accompanies the presence of a refinery. Aside from air pollution impacts there are also wastewater concerns, risks of industrial accidents such as fire and explosion, and noise health effects due to industrial noise.

Many governments worldwide have mandated restrictions on contaminants that refineries release, and most refineries have installed the equipment needed to comply with the requirements of the pertinent environmental protection regulatory agencies. In the United States, there is strong pressure to prevent the development of new refineries, and no major refinery has been built in the country since Marathon'sGaryville, Louisiana facility in 1976. However, many existing refineries have been expanded during that time. Environmental restrictions and pressure to prevent construction of new refineries may have also contributed to rising fuel prices in the United States. Additionally, many refineries (more than 100 since the 1980s) have closed due to obsolescence and/or merger activity within the industry itself.

Environmental and safety concerns mean that oil refineries are sometimes located some distance away from major urban areas. Nevertheless, there are many instances where refinery operations are close to populated areas and pose health risks. In California's Contra Costa County and Solano County, a shoreline necklace of refineries, built in the early 20th century before this area was populated, and associated chemical plants are adjacent to urban areas in Richmond, Martinez, Pacheco, Concord, Pittsburg, Vallejo and Benicia, with occasional accidental events that require "shelter in place" orders to the adjacent populations.

NIOSH criteria for occupational exposure to refined petroleum solvents have been available since 1977.

Corrosion Problems and Prevention

Refinery of Slovnaft in Bratislava.

Petroleum refineries run as efficiently as possible to reduce costs. One major factor that decreases efficiency is corrosion of the metallic components found throughout refining process. Corrosion causes the failure of equipment items as well as dictating the maintenance schedule of the refinery, during which part or all of the refinery must be shut down. The corrosion-related direct costs in the U.S. petroleum industry as of 1996 was estimated as US$3.7 billion per year.

Corrosion occurs in various forms in the refining process, such as pitting corrosion from water droplets, embrittlement from hydrogen, and stress corrosion cracking from sulfide attack. From a materials standpoint, carbon steel is used for upwards of 80 per cent of refinery components, which is beneficial due to its low cost. Carbon steel is resistant to the most common forms of corrosion, particularly from hydrocarbon impurities at temperatures below 205 °C, but other corrosive chemicals and environments prevent its use everywhere. Common replacement materials are low alloy steels containing chromium and molybdenum, with stainless steels containing more chromium dealing with more corrosive environments. More expensive materials commonly used are nickel, titanium, and copper alloys. These are primarily saved for the most problematic areas where extremely high temperatures and/or very corrosive chemicals are present.

Corrosion is fought by a complex system of monitoring, preventative repairs and careful use of materials. Monitoring methods include both off-line checks taken during maintenance and on-line monitoring. Off-line checks measure corrosion after it has occurred, telling the engineer when equipment must be replaced based on the historical information he has collected. This is referred to as preventative management.

Oil refinery in Iran.

On-line systems are a more modern development, and are revolutionizing the way corrosion is approached. There are several types of on-line corrosion monitoring technologies such as linear polarization resistance, electrochemical noise and electrical resistance. On-Line monitoring has generally had slow reporting rates in the past (minutes or hours) and been limited by process conditions and sources of error but newer technologies can report rates up to twice per minute with much higher accuracy (referred to as real-time monitoring). This allows process engineers to treat corrosion as another process variable that can be optimized in the system. Immediate responses to process changes allow the control of corrosion mechanisms, so they can be minimized while also maximizing production output. In an ideal situation having on-line corrosion information that is accurate and real-time will allow conditions that cause high corrosion rates to be identified and reduced. This is known as predictive management.

Materials methods include selecting the proper material for the application. In areas of minimal corrosion, cheap materials are preferable, but when bad corrosion can occur, more expensive but longer lasting materials should be used. Other materials methods come in the form of protective barriers between corrosive substances and the equipment metals. These can be either a lining of refractory material such as standard Portland cement or other special acid-resistant cements that are shot onto the inner surface of the vessel. Also available are thin overlays of more expensive metals that protect cheaper metal against corrosion without requiring lots of material.

History

The world's first refinery opened at Ploie□ti, Romania, in 1856-1857. After being taken over by Nazi Germany, the Ploie□ti refineries were bombed in Operation Tidal Wave by the Allies during the Oil Campaign of World War II.

At one point, the refinery in Ras Tanura, Saudi Arabia owned by Saudi Aramco was claimed to be the largest oil refinery in the world. For most of the 20th century, the largest refinery was the Abadan Refinery in Iran. This refinery suffered extensive damage during the Iran–Iraq War. On the 31 December 2014, the world's largest refinery complex is the Jamnagar Refinery Complex, consisting of two refineries side by side operated by Reliance Industries Limited in Jamnagar, India with a combined production capacity of 1,240,000 barrels per day (197,000 m^3/d). PDVSA's Paraguaná Refinery Complex in Paraguaná Peninsula, Venezuela with a capacity of 940,000 bbl/d (149,000 m^3/d) and SK Energy's Ulsan in South Korea with 840,000 bbl/d (134,000 m^3/d) are the second and third largest, respectively.

Oil Refining in the United States

In the 19th century, refineries in the U.S. processed crude oil primarily to recover the kerosene. There was no market for the more volatile fraction, including gasoline, which was considered waste and was often dumped directly into the nearest river. The invention of the automobile shifted the demand to gasoline and diesel, which remain the primary refined products today. Today, national and state legislation requires refineries to meet stringent air and water cleanliness standards. In fact, oil companies in the U.S. perceive obtaining a permit to build a modern refinery to be so difficult and costly that no new refineries were built (though many have been expanded) in the U.S. from 1976 until 2014, when the small Dakota Prairie Refinery in North Dakota is set to begin operation. More than half the refineries that existed in 1981 are now closed due to low utilization rates and accelerating mergers. As a result of these closures total US refinery capacity fell between 1981 and 1995, though the operating capacity stayed fairly constant in that time period at around 15,000,000 barrels per day (2,400,000 m^3/d). Increases in facility size and improvements in efficiencies have offset much of the lost physical capacity of the industry. In 1982 (the earliest data provided), the United States operated 301 refineries with a combined capacity of 17.9 million barrels (2,850,000 m^3) of crude oil each calendar day. In 2010, there were 149 operable U.S. refineries with a combined capacity of 17.6 million barrels (2,800,000 m^3) per calendar day. By 2014 the number of refinery had reduced to 140 but the total capacity increased to 18.02 million barrels (2,865,000 m^3) per calendar day. Indeed, in order to reduce operating costs and depreciation, refining is operated in less sites but of bigger capacity.

In 2009 through 2010, as revenue streams in the oil business dried up and profitability of oil refineries fell due to lower demand for product and high reserves of supply preceding the economic recession, oil companies began to close or sell the less profitable refineries.

ExxonMobil oil refinery in Baton Rouge, Louisiana (the fourth-largest in the United States).

Worldwide Oil Refining Capacity

According to the Oil and Gas Journal in the world a total of 636 refineries were operated on the 31 December 2014 for a total capacity of 87.75 million barrels (13,951,000 m^3).

Amine Gas Treating

Amine gas treating, also known as amine scrubbing, gas sweetening and acid gas removal, refers to a group of processes that use aqueous solutions of various alkylamines (commonly referred to simply as amines) to remove hydrogen sulfide (H_2S) and carbon dioxide (CO_2) from gases. It is a common unit process used in refineries, and is also used in petrochemical plants, natural gas processing plants and other industries.

Processes within oil refineries or chemical processing plants that remove hydrogen sulfide are referred to as "sweetening" processes because the odor of the processed products is improved by the absence of hydrogen sulfide. An alternative to the use of amines involves membrane technology. However, membrane separation is less attractive due to the relatively high capital and operating costs as well as other technical factors.

Many different amines are used in gas treating:

- Diethanolamine (DEA).
- Monoethanolamine (MEA).
- Methyldiethanolamine (MDEA).
- Diisopropanolamine (DIPA).
- Aminoethoxyethanol (Diglycolamine) (DGA).

The most commonly used amines in industrial plants are the alkanolamines DEA, MEA, and MDEA. These amines are also used in many oil refineries to remove sour gases from liquid hydrocarbons such as liquified petroleum gas (LPG).

Description of a Typical Amine Treater

Gases containing H_2S or both H_2S and CO_2 are commonly referred to as sour gases or acid gases in the hydrocarbon processing industries.

The chemistry involved in the amine treating of such gases varies somewhat with the particular amine being used. For one of the more common amines, monoethanolamine (MEA) denoted as RNH_2, the chemistry may be expressed as:

$$RNH_2 + H_2S \Leftrightarrow RNH+3 + SH^-$$

A typical amine gas treating process (the Girbotol process, as shown in the flow diagram below) includes an absorber unit and a regenerator unit as well as accessory equipment. In the absorber, the downflowing amine solution absorbs H_2S and CO_2 from the upflowing sour gas to produce a sweetened gas stream (i.e., a gas free of hydrogen sulfide and carbon dioxide) as a product and an amine solution rich in the absorbed acid gases. The resultant "rich" amine is then routed into the regenerator (a stripper with a reboiler) to produce regenerated or "lean" amine that is recycled for reuse in the absorber. The stripped overhead gas from the regenerator is concentrated H_2S and CO_2.

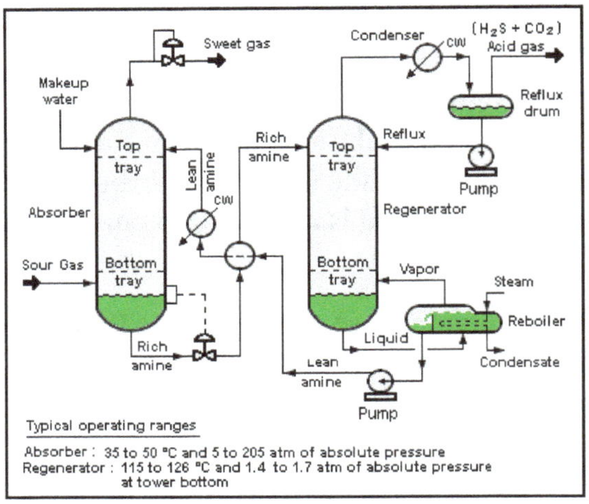

Process flow diagram of a typical amine treating process used in petroleum refineries, natural gas processing plants and other industrial facilities.

Alternative Processes

Alternative stripper configurations include matrix, internal exchange, flashing feed, and multi-pressure with split feed. Many of these configurations offer more energy efficiency for specific solvents or operating conditions. Vacuum operation favors solvents with low heats of absorption while operation at normal pressure favors solvents with high heats of absorption. Solvents with high heats of absorption require less energy for stripping from temperature swing at fixed capacity. The matrix stripper recovers 40% of CO_2 at a higher pressure and does not have inefficiencies associated with multipressure stripper. Energy and costs are reduced since the reboiler duty cycle is slightly less than normal pressure stripper. An Internal Exchange stripper has a smaller ratio of water vapor to CO_2 in the overheads stream, and therefore less steam is required. The multi-pressure configuration with split feed reduces the flow into the bottom section, which also reduces the equivalent work. Flashing feed requires less heat input because it uses the latent heat of water vapor to help strip some of the CO_2 in the rich stream entering the stripper at the bottom of the column. The multipressure configuration is more attractive for solvents with a higher heats of absorption.

Amines

The amine concentration in the absorbent aqueous solution is an important parameter in the design and operation of an amine gas treating process. Depending on which one of the following four amines the unit was designed to use and what gases it was designed to remove, these are some typical amine concentrations, expressed as weight percent of pure amine in the aqueous solution:

- Monoethanolamine: About 20 % for removing H_2S and CO_2, and about 32 % for removing only CO_2.
- Diethanolamine: About 20 to 25 % for removing H_2S and CO_2.
- Methyldiethanolamine: About 30 to 55% % for removing H_2S and CO_2.
- Diglycolamine: About 50 % for removing H_2S and CO_2.

The choice of amine concentration in the circulating aqueous solution depends upon a number of factors and may be quite arbitrary. It is usually made simply on the basis of experience. The factors involved include whether the amine unit is treating raw natural gas or petroleum refinery by-product gases that contain relatively low concentrations of both H_2S and CO_2 or whether the unit is treating gases with a high percentage of CO_2 such as the offgas from the steam reforming process used in ammonia production or the flue gases from power plants.

Both H_2S and CO_2 are acid gases and hence corrosive to carbon steel. However, in an amine treating unit, CO_2 is the stronger acid of the two. H_2S forms a film of iron sulfide on the surface of the steel that acts to protect the steel. When treating gases with a high percentage of CO_2, corrosion inhibitors are often used and that permits the use of higher concentrations of amine in the circulating solution.

Another factor involved in choosing an amine concentration is the relative solubility of H_2S and CO_2 in the selected amine. The choice of the type of amine will affect the required circulation rate of amine solution, the energy consumption for the regeneration and the ability to selectively remove either H_2S alone or CO_2 alone if desired. For more information about selecting the amine concentration, the reader is referred to Kohl and Nielsen's book.

MEA and DEA

MEA and DEA are primary and secondary amines. They are very reactive and can effectively remove a high volume of gas removal due to a high reaction rate. However, due to stoichiometry, the loading capacity is limited to 0.5 mol CO_2 per mole of amine. MEA and DEA also require a large amount of energy to strip the CO_2 to during regeneration, which can be up to 70% of total operating costs. They are also more corrosive and chemically unstable compared to other amines.

Uses

In oil refineries, that stripped gas is mostly H_2S, much of which often comes from a sulfur-removing process called hydrodesulfurization. This H_2S-rich stripped gas stream is then usually routed into a Claus process to convert it into elemental sulfur. In fact, the vast majority of the 64,000,000 metric tons of sulfur produced worldwide in 2005 was byproduct sulfur from refineries and other hydrocarbon processing plants. Another sulfur-removing process is the WSA Process which recovers sulfur in any form as concentrated sulfuric acid. In some plants, more than one amine absorber unit may share a common regenerator unit. The current emphasis on removing CO_2 from the flue gases emitted by fossil fuel power plants has led to much interest in using amines for removing CO_2.

In the specific case of the industrial synthesis of ammonia, for the steam reforming process of hydrocarbons to produce gaseous hydrogen, amine treating is one of the commonly used processes for removing excess carbon dioxide in the final purification of the gaseous hydrogen.

In the biogas production it is sometimes necessary to remove carbon dioxide from the biogas to make it comparable with the natural. The removal of the sometimes high content of hydrogen sulfide is necessary to prevent corrosion of metallic parts after burning the bio gas.

Carbon Capture and Storage

Amines are used to remove CO_2 in various areas ranging from natural gas production to the food and beverage industry, and have been for over sixty years.

There are multiple classifications of amines, each of which has different characteristics relevant to CO_2 capture. For example, Monoethanolamine (MEA) reacts strongly with acid gases like CO_2 and has a fast reaction time and an ability to remove high percentages of CO_2, even at the low CO_2 concentrations. Typically, Monoethanolamine (MEA) can capture 85% to 90% of the CO_2 from the flue gas of a coal-fired plant, which is one of the most effective solvent to capture CO_2.

Challenges of carbon capture using amine include:

- Low pressure gas increases difficulty of transferring CO_2 from the gas into amine.
- Oxygen content of the gas can cause amine degradation and acid formation.
- CO_2 degradation of primary (and secondary) amines.
- High energy consumption.
- Very large facilities.
- Finding suitable location for the removed CO_2.

The partial pressure is the driving force to transfer CO_2 into the liquid phase. Under the low pressure, this transfer is hard to achieve without increasing the reboiler's heat duty, which will result in higher cost.

Primary and secondary amines, for example, MEA and DEA, will react with CO_2 and form degradation products. O_2 from the inlet gas will cause degradation as well. The degraded amine is no longer able to capture CO_2, which decreases the overall carbon capture efficiency.

Currently, variety of amine mixtures are being synthesized and tested to achieve a more desirable set of overall properties for use in CO_2 capture systems. One major focus is on lowering the energy required for solvent regeneration, which has a major impact on process costs. However, there are tradeoffs to consider. For example, the energy required for regeneration is typically related to the driving forces for achieving high capture capacities. Thus, reducing the regeneration energy can lower the driving force and thereby increase the amount of solvent and size of absorber needed to capture a given amount of CO_2, thus, increasing the capital cost.

Catalytic Reforming

Catalytic reforming is a chemical process used to convert petroleum refinerynaphthas distilled from crude oil (typically having low octane ratings) into high-octane liquid products called reformates, which are premium blending stocks for high-octane gasoline. The process converts low-octane linear hydrocarbons (paraffins) into branched alkanes (isoparaffins) and cyclic naphthenes, which are then partially dehydrogenated to produce high-octane aromatic hydrocarbons. The dehydrogenation also produces significant amounts of byproduct hydrogen gas, which is fed into other refinery processes such as hydrocracking. A side reaction is hydrogenolysis, which produces light hydrocarbons of lower value, such as methane, ethane, propane and butanes. It is also the conversion of straight chains of alkane catalytically.

In addition to a gasoline blending stock, reformate is the main source of aromatic bulk chemicals such as benzene, toluene, xylene and ethylbenzene which have diverse uses, most importantly as raw materials for conversion into plastics. However, the benzene content of reformate makes it carcinogenic, which has led to governmental regulations effectively requiring further processing to reduce its benzene content.

This process is quite different from and not to be confused with the catalytic steam reforming process used industrially to produce products such as hydrogen, ammonia, and methanol from natural gas, naphtha or other petroleum-derived feedstocks. Nor is this process to be confused with various other catalytic reforming processes that use methanol or biomass-derived feedstocks to produce hydrogen for fuel cells or other uses.

History

In the 1940s, Vladimir Haensel, a research chemist working for Universal Oil Products (UOP), developed a catalytic reforming process using a catalyst containing platinum. Haensel's process was subsequently commercialized by UOP in 1949 for producing a high octane gasoline from low octane naphthas and the UOP process become known as the Platforming process. The first Platforming unit was built in 1949 at the refinery of the Old Dutch Refining Company in Muskegon, Michigan.

In the years since then, many other versions of the process have been developed by some of the major oil companies and other organizations. Today, the large majority of gasoline produced worldwide is derived from the catalytic reforming process.

To name a few of the other catalytic reforming versions that were developed, all of which utilized a platinum and/or a rhenium catalyst:

- Rheniforming: Developed by Chevron Oil Company.
- Powerforming: Developed by Esso Oil Company, currently known as ExxonMobil.
- Magnaforming: Developed by Engelhard and Atlantic Richfield Oil Company.
- Ultraforming: Developed by Standard Oil of Indiana, now a part of the British Petroleum Company.
- Houdriforming: Developed by the Houdry Process Corporation.
- CCR Platforming: A Platforming version, designed for continuous catalyst regeneration, developed by UOP.

- Octanizing: A catalytic reforming version developed by Axens, a subsidiary of Institut francais du petrole (IFP), designed for continuous catalyst regeneration.

Chemistry

Before describing the reaction chemistry of the catalytic reforming process as used in petroleum refineries, the typical naphthas used as catalytic reforming feedstocks will be discussed.

Typical Naphtha Feedstocks

A petroleum refinery includes many unit operations and unit processes. The first unit operation in a refinery is the continuous distillation of the petroleum crude oil being refined. The overhead liquid distillate is called naphtha and will become a major component of the refinery's gasoline (petrol) product after it is further processed through a catalytic hydrodesulfurizer to remove sulfur-containing hydrocarbons and a catalytic reformer to reform its hydrocarbon molecules into more complex molecules with a higher octane rating value. The naphtha is a mixture of very many different hydrocarbon compounds. It has an initial boiling point of about 35 °C and a final boiling point of about 200 °C, and it contains paraffin, naphthene (cyclic paraffins) and aromatic hydrocarbons ranging from those containing 4 carbon atoms to those containing about 10 or 11 carbon atoms.

The naphtha from the crude oil distillation is often further distilled to produce a "light" naphtha containing most (but not all) of the hydrocarbons with 6 or fewer carbon atoms and a "heavy" naphtha containing most (but not all) of the hydrocarbons with more than 6 carbon atoms. The heavy naphtha has an initial boiling point of about 140 to 150 °C and a final boiling point of about 190 to 205 °C. The naphthas derived from the distillation of crude oils are referred to as "straight-run" naphthas.

It is the straight-run heavy naphtha that is usually processed in a catalytic reformer because the light naphtha has molecules with 6 or fewer carbon atoms which, when reformed, tend to crack into butane and lower molecular weight hydrocarbons which are not useful as high-octane gasoline blending components. Also, the molecules with 6 carbon atoms tend to form aromatics which is undesirable because governmental environmental regulations in a number of countries limit the amount of aromatics (most particularly benzene) that gasoline may contain.

It should be noted that there are a great many petroleum crude oil sources worldwide and each crude oil has its own unique composition or "assay". Also, not all refineries process the same crude oils and each refinery produces its own straight-run naphthas with their own unique initial and final boiling points. In other words, naphtha is a generic term rather than a specific term.

The table just below lists some fairly typical straight-run heavy naphtha feedstocks, available for catalytic reforming, derived from various crude oils. It can be seen that they differ significantly in their content of paraffins, naphthenes and aromatics:

Typical Heavy Naphtha Feedstocks				
Crude oil name Location	**Barrow Island Australia**	**Mutineer-Exeter Australia**	**CPC Blend Kazakhstan**	**Draugen North Sea**
Initial boiling point, °C	149	140	149	150
Final boiling point, °C	204	190	204	180

Paraffins, liquid volume %	46	62	57	38
Naphthenes, liquid volume %	42	32	27	45
Aromatics, liquid volume %	12	6	16	17

Some refinery naphthas include olefinic hydrocarbons, such as naphthas derived from the fluid catalytic cracking and coking processes used in many refineries. Some refineries may also desulfurize and catalytically reform those naphthas. However, for the most part, catalytic reforming is mainly used on the straight-run heavy naphthas, such as those in the above table, derived from the distillation of crude oils.

The Reaction Chemistry

There are many chemical reactions that occur in the catalytic reforming process, all of which occur in the presence of a catalyst and a high partial pressure of hydrogen. Depending upon the type or version of catalytic reforming used as well as the desired reaction severity, the reaction conditions range from temperatures of about 495 to 525 °C and from pressures of about 5 to 45 atm.

The commonly used catalytic reforming catalysts contain noble metals such as platinum and/or rhenium, which are very susceptible to poisoning by sulfur and nitrogen compounds. Therefore, the naphtha feedstock to a catalytic reformer is always pre-processed in a hydrodesulfurization unit which removes both the sulfur and the nitrogen compounds. Most catalysts require both sulphur and nitrogen content to be lower than 1 ppm.

The four major catalytic reforming reactions are:

1: The dehydrogenation of naphthenes to convert them into aromatics as exemplified in the conversion methylcyclohexane (a naphthene) to toluene (an aromatic), as shown below:

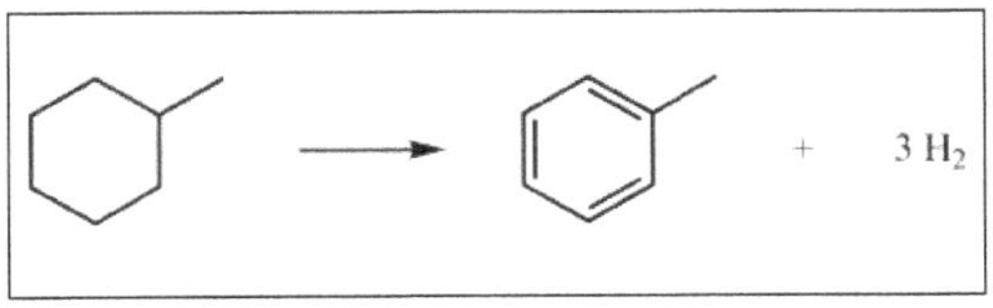

2: The isomerization of normal paraffins to isoparaffins as exemplified in the conversion of normal octane to 2,5-Dimethylhexane (an isoparaffin), as shown below:

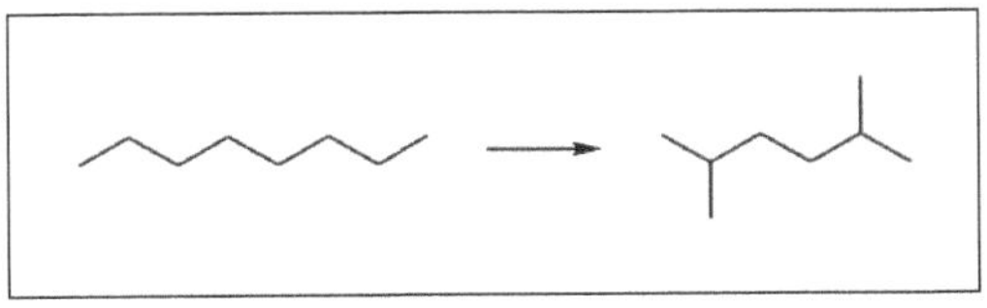

3: The dehydrogenation and aromatization of paraffins to aromatics (commonly called dehydrocyclization) as exemplified in the conversion of normal heptane to toluene, as shown below:

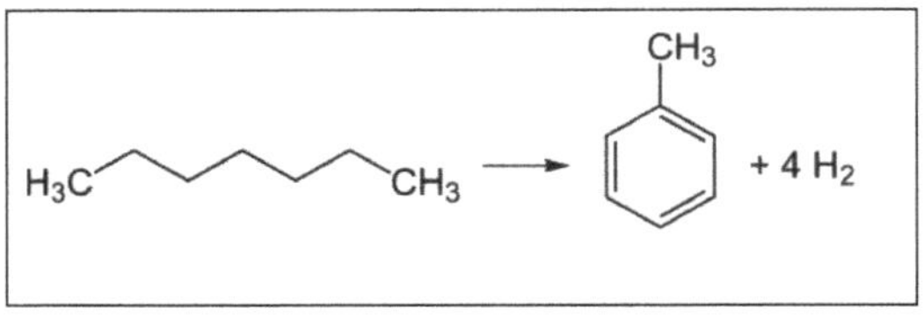

4: The hydrocracking of paraffins into smaller molecules as exemplified by the cracking of normal heptane into isopentane and ethane, as shown below:

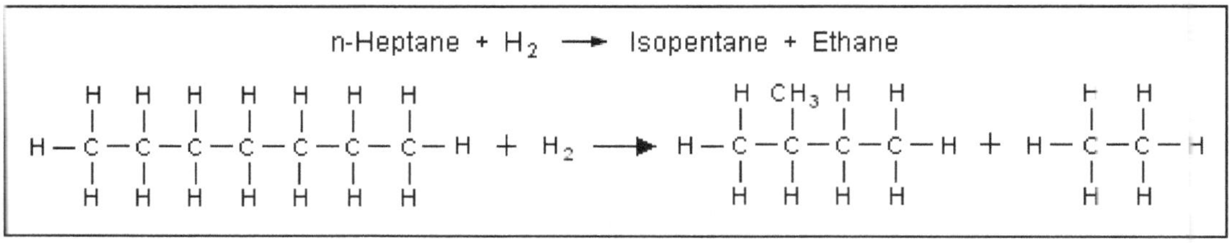

The hydrocracking of paraffins is the only one of the above four major reforming reactions that consumes hydrogen. The isomerization of normal paraffins does not consume or produce hydrogen. However, both the dehydrogenation of naphthenes and the dehydrocyclization of paraffins produce hydrogen. The overall net production of hydrogen in the catalytic reforming of petroleum naphthas ranges from about 50 to 200 cubic meters of hydrogen gas (at 0 °C and 1 atm) per cubic meter of liquid naphtha feedstock. In the United States customary units, that is equivalent to 300 to 1200 cubic feet of hydrogen gas (at 60 °F and 1 atm) per barrel of liquid naphtha feedstock. In many petroleum refineries, the net hydrogen produced in catalytic reforming supplies a significant part of the hydrogen used elsewhere in the refinery (for example, in hydrodesulfurization processes). The hydrogen is also necessary in order to hydrogenolyze any polymers that form on the catalyst.

In practice, the higher the content of naphtenes in the naphtha feedstock, the better will be the quality of the reformate and the higher the production of hydrogen. Crude oils containing the best naphtha for reforming are typically from Western Africa or the North Sea, such as Bonny light or Troll.

Process Description

The most commonly used type of catalytic reforming unit has three reactors, each with a fixed bed of catalyst, and all of the catalyst is regenerated in situ during routine catalyst regeneration shutdowns which occur approximately once each 6 to 24 months. Such a unit is referred to as a semi-regenerative catalytic reformer (SRR).

Some catalytic reforming units have an extra *spare* or *swing* reactor and each reactor can be individually isolated so that any one reactor can be undergoing in situ regeneration while the other reactors are in operation. When that reactor is regenerated, it replaces another reactor which, in turn, is isolated so that it can then be regenerated. Such units, referred to as *cyclic* catalytic reformers, are not very common. Cyclic catalytic reformers serve to extend the period between required shutdowns.

The latest and most modern type of catalytic reformers are called continuous catalyst regeneration (CCR) reformers. Such units are characterized by continuous in-situ regeneration of part of the catalyst in a special regenerator, and by continuous addition of the regenerated catalyst to the operating reactors. As of 2006, two CCR versions available: UOP's CCR Platformer process and Axens' Octanizing process. The installation and use of CCR units is rapidly increasing.

Many of the earliest catalytic reforming units (in the 1950s and 1960s) were non-regenerative in that they did not perform in situ catalyst regeneration. Instead, when needed, the aged catalyst was replaced by fresh catalyst and the aged catalyst was shipped to catalyst manufacturers to be either regenerated or to recover the platinum content of the aged catalyst. Very few, if any, catalytic reformers currently in operation are non-regenerative.

The process flow diagram below depicts a typical semi-regenerative catalytic reforming unit.

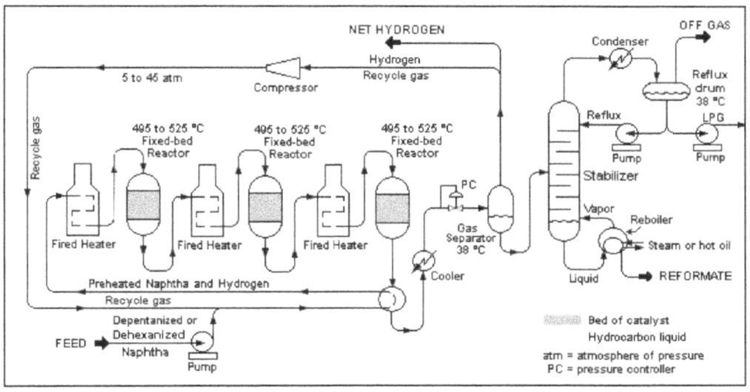

Schematic diagram of a typical semi-regenerative catalytic reformer unit in a petroleum refinery.

The liquid feed (at the bottom left in the diagram) is pumped up to the reaction pressure (5–45 atm) and is joined by a stream of hydrogen-rich recycle gas. The resulting liquid–gas mixture is preheated by flowing through a heat exchanger. The preheated feed mixture is then totally vaporized and heated to the reaction temperature (495–520 °C) before the vaporized reactants enter the first reactor. As the vaporized reactants flow through the fixed bed of catalyst in the reactor, the major reaction is the dehydrogenation of naphthenes to aromatics (as described earlier herein) which is highly endothermic and results in a large temperature decrease between the inlet and outlet of the reactor. To maintain the required reaction temperature and the rate of reaction, the vaporized stream is reheated in the second fired heater before it flows through the second reactor. The temperature again decreases across the second reactor and the vaporized stream must again be reheated in the third fired heater before it flows through the third reactor. As the vaporized stream proceeds through the three reactors, the reaction rates decrease and the reactors therefore become larger. At the same time, the amount of reheat required between the reactors becomes smaller. Usually, three reactors are all that is required to provide the desired performance of the catalytic reforming unit.

Some installations use three separate fired heaters as shown in the schematic diagram and some installations use a single fired heater with three separate heating coils.

The hot reaction products from the third reactor are partially cooled by flowing through the heat exchanger where the feed to the first reactor is preheated and then flow through a water-cooled heat exchanger before flowing through the pressure controller (PC) into the gas separator.

Most of the hydrogen-rich gas from the gas separator vessel returns to the suction of the recycle hydrogen gas compressor and the net production of hydrogen-rich gas from the reforming reactions is exported for use in the other refinery processes that consume hydrogen (such as hydrodesulfurization units and/or a hydrocracker unit).

The liquid from the gas separator vessel is routed into a fractionating column commonly called a *stabilizer*. The overhead offgas product from the stabilizer contains the byproduct methane, ethane, propane and butane gases produced by the hydrocracking reactions as explained in the above discussion of the reaction chemistry of a catalytic reformer, and it may also contain some small amount of hydrogen. That offgas is routed to the refinery's central gas processing plant for removal and recovery of propane and butane. The residual gas after such processing becomes part of the refinery's fuel gas system.

The bottoms product from the stabilizer is the high-octane liquid reformate that will become a component of the refinery's product gasoline. Reformate can be blended directly in the gasoline pool but often it is separated in two or more streams. A common refining scheme consists in fractionating the reformate in two streams, light and heavy reformate. The light reformate has lower octane and can be used as isomerization feedstock if this unit is available. The heavy reformate is high in octane and low in benzene, hence it is an excellent blending component for the gasoline pool.

Benzene is often removed with a specific operation to reduce the content of benzene in the reformate as the finished gasoline has often an upper limit of benzene content (in the UE this is 1% volume). The benzene extracted can be marketed as feedstock for the chemical industry.

Catalysts and Mechanisms

Most catalytic reforming catalysts contain platinum or rhenium on a silica or silica-alumina support base, and some contain both platinum and rhenium. Fresh catalyst is chlorided (chlorinated) prior to use.

The noble metals (platinum and rhenium) are considered to be catalytic sites for the dehydrogenation reactions and the chlorinated alumina provides the acid sites needed for isomerization, cyclization and hydrocracking reactions. The biggest care has to be exercised during the chlorination. Indeed, if not chlorinated (or insufficiently chlorinated) the platinum and rhenium in the catalyst would be reduced almost immediately to metallic state by the hydrogen in the vapour phase. On the other an excessive chlorination could depress excessively the activity of the catalyst.

The activity (i.e., effectiveness) of the catalyst in a semi-regenerative catalytic reformer is reduced over time during operation by carbonaceous coke deposition and chloride loss. The activity of the catalyst can be periodically regenerated or restored by in situ high temperature oxidation of the coke followed by chlorination. As stated earlier herein, semi-regenerative catalytic reformers are regenerated about once per 6 to 24 months. The higher the severity of the reacting conditions (temperature), the higher is the octane of the produced reformate but also the shorter will be the duration of the cycle between two regenerations. Catalyst's cycle duration is also very dependent on the quality of the feedstock. However, independently of the crude oil used in the refinery, all catalysts require a maximum final boiling point of the naphtha feedstock of 180 °C.

Normally, the catalyst can be regenerated perhaps 3 or 4 times before it must be returned to the manufacturer for reclamation of the valuable platinum and/or rhenium content.

Cracking

In petroleum geology and chemistry, cracking is the process whereby complex organicmolecules such as kerogens or long chain hydrocarbons are broken down into simpler molecules such as light hydrocarbons, by the breaking of carbon-carbon bonds in the precursors. The rate of cracking and the end products are strongly dependent on the temperature and presence of catalysts. Cracking is the breakdown of a large alkane into smaller, more useful alkanes and alkenes. Simply put, hydrocarbon cracking is the process of breaking a long-chain of hydrocarbons into short ones. This process might require high temperatures and high pressure.

More loosely, outside the field of petroleum chemistry, the term "cracking" is used to describe any type of splitting of molecules under the influence of heat, catalysts and solvents, such as in processes of destructive distillation or pyrolysis.

Fluid catalytic cracking produces a high yield of petrol and LPG, while hydrocracking is a major source of jet fuel, Diesel fuel, naphtha, and again yields LPG.

Refinery using the Shukhov cracking process, Baku, Soviet Union, 1934.

History and Patents

Among several variants of thermal cracking methods (variously known as the "Shukhov cracking process", "Burton cracking process", "Burton-Humphreys cracking process", and "Dubbs cracking process") Vladimir Shukhov, a Russian engineer, invented and patented the first in 1891 (Russian Empire, patent no. 12926, November 27, 1891). One installation was used to a limited extent in Russia, but development was not followed up. In the first decade of the 20th century the American engineers William Merriam Burton and Robert E. Humphreys independently developed and patented a similar process as U.S. patent 1,049,667 on June 8, 1908. Among its advantages was the fact that both the condenser and the boiler were continuously kept under pressure.

In its earlier versions however, it was a batch process, rather than continuous, and many patents were to follow in the USA and Europe, though not all were practical. In 1924, a delegation from the American Sinclair Oil Corporation visited Shukhov. Sinclair Oil apparently wished to suggest that the patent of Burton and Humphreys, in use by Standard Oil, was derived from Shukhov's patent for oil cracking, as described in the Russian patent. If that could be established, it could strengthen the hand of rival American companies wishing to invalidate the Burton-Humphreys patent. In the event Shukhov satisfied the Americans that in principle Burton's method closely resembled his 1891 patents, though his own interest in the matter was primarily to establish that "the Russian oil industry could easily build a cracking apparatus according to any of the described systems without being accused by the Americans of borrowing for free".

At that time, just a few years after the Russian Revolution, Russia was desperate to develop industry and earn foreign exchange, so their oil industry eventually did obtain much of their technology from foreign companies, largely American. At about that time however, fluid catalytic cracking was being explored and developed and soon replaced most of the purely thermal cracking processes in the fossil fuel processing industry. The replacement was however not complete; many types of cracking, including pure thermal cracking, still are in use, depending on the nature of the feedstock and the products required to satisfy market demands. Thermal cracking remains important however, for example in producing naphtha, gas oil, and coke, and more sophisticated forms of thermal cracking have been developed for various purposes. These include visbreaking, steam cracking, and coking.

Chemistry

A large number of chemical reactions take place during the cracking process, most of them based on free radicals. Computer simulations aimed at modeling what takes place during steam cracking have included hundreds or even thousands of reactions in their models. The main reactions that take place include:

Initiation

In these reactions a single molecule breaks apart into two free radicals. Only a small fraction of the feed molecules actually undergo initiation, but these reactions are necessary to produce the free radicals that drive the rest of the reactions. In steam cracking, initiation usually involves breaking a chemical bond between two carbon atoms, rather than the bond between a carbon and a hydrogen atom.

$$CH_3CH_3 \rightarrow 2\ CH_3\bullet$$

Hydrogen Abstraction

In these reactions a free radical removes a hydrogen atom from another molecule, turning the second molecule into a free radical.

$$CH_3\bullet + CH_3CH_3 \rightarrow CH_4 + CH_3CH_2\bullet$$

Radical Decomposition

In these reactions a free radical breaks apart into two molecules, one an alkene, the other a free radical. This is the process that results in alkene products.

$$CH_3CH_2\bullet \rightarrow CH_2{=}CH_2 + H$$

Radical Addition

In these reactions, the reverse of radical decomposition reactions, a radical reacts with an alkene to form a single, larger free radical. These processes are involved in forming the aromatic products that result when heavier feedstocks are used.

$$CH_3CH_2\bullet + CH_2{=}CH_2 \rightarrow CH_3CH_2CH_2CH_2\bullet$$

Termination

In these reactions two free radicals react with each other to produce products that are not free radicals. Two common forms of termination are *recombination*, where the two radicals combine to form one larger molecule, and *disproportionation*, where one radical transfers a hydrogen atom to the other, giving an alkene and an alkane.

$$CH_3\bullet + CH_3CH_2\bullet \rightarrow CH_3CH_2CH_3$$

$$CH_3CH_2\bullet + CH_3CH_2\bullet \rightarrow CH_2{=}CH_2 + CH_3CH_3$$

Example: Cracking Butane

There are three places where a butane molecule (CH_3-CH_2-CH_2-CH_3) might be split. Each has a distinct likelihood:

- 48%: break at the CH_3-CH_2 bond.

  $$CH_3{*} \,/\, {*}CH_2\text{-}CH_2\text{-}CH_3$$

  Ultimately this produces an alkane and an alkene: $CH_4 + CH_2{=}CH\text{-}CH_3$

- 38%: break at a CH_2-CH_2 bond.

  $$CH_3\text{-}CH_2{*} \,/\, {*}CH_2\text{-}CH_3$$

  Ultimately this produces an alkane and an alkene of different types: $CH_3\text{-}CH_3 + CH_2{=}CH_2$

- 14%: break at a terminal C-H bond.

  $$H/CH_2\text{-}CH_2\text{-}CH_2\text{-}CH_3$$

  Ultimately this produces an alkene and hydrogen gas: $CH_2{=}CH\text{-}CH_2\text{-}CH_3 + H_2$

Cracking Methodologies

Thermal Methods

Thermal cracking was the first category of hydrocarbon cracking to be developed. Thermal cracking is an example of a reaction whose energetics are dominated by entropy ($\Delta S°$) rather than by enthalpy ($\Delta H°$) in the Gibbs Free Energy equation $\Delta G° = \Delta H° - T\Delta S°$. Although the bond dissociation energy D for a carbon-carbon single bond is relatively high (about 375 kJ/mol) and cracking is highly endothermic, the large positive entropy change resulting from the fragmentation of one large molecule into several smaller pieces, together with the extremely high temperature, makes $T\Delta S^{\circ}$ term larger than the ΔH° term, thereby favoring the cracking reaction.

Thermal Cracking

Modern high-pressure thermal cracking operates at absolute pressures of about 7,000 kPa. An overall process of disproportionation can be observed, where "light", hydrogen-rich products are formed at the expense of heavier molecules which condense and are depleted of hydrogen. The actual reaction is known as homolytic fission and produces alkenes, which are the basis for the economically important production of polymers.

Thermal cracking is currently used to "upgrade" very heavy fractions or to produce light fractions or distillates, burner fuel and/or petroleum coke. Two extremes of the thermal cracking in terms of product range are represented by the high-temperature process called "steam cracking" or pyrolysis (ca. 750 °C to 900 °C or higher) which produces valuable ethylene and other feedstocks for the petrochemical industry, and the milder-temperature delayed coking (ca. 500 °C) which can produce, under the right conditions, valuable needle coke, a highly crystalline petroleum coke used in the production of electrodes for the steel and aluminium industries.

William Merriam Burton developed one of the earliest thermal cracking processes in 1912 which operated at 700–750 °F (371–399 °C) and an absolute pressure of 90 psi (620 kPa) and was known as the Burton process. Shortly thereafter, in 1921, C.P. Dubbs, an employee of the Universal Oil Products Company, developed a somewhat more advanced thermal cracking process which operated at 750–860 °F (399–460 °C) and was known as the *Dubbs process*. The Dubbs process was used extensively by many refineries until the early 1940s when catalytic cracking came into use.

Steam Cracking

Steam cracking is a petrochemical process in which saturated hydrocarbons are broken down into smaller, often unsaturated, hydrocarbons. It is the principal industrial method for producing the lighter alkenes (or commonly olefins), including ethene (or ethylene) and propene (or propylene). Steam cracker units are facilities in which a feedstock such as naphtha, liquefied petroleum gas (LPG), ethane, propane or butane is thermally cracked through the use of steam in a bank of pyrolysis furnaces to produce lighter hydrocarbons. The products obtained depend on the composition of the feed, the hydrocarbon-to-steam ratio, and on the cracking temperature and furnace residence time.

In steam cracking, a gaseous or liquid hydrocarbon feed like naphtha, LPG or ethane is diluted with steam and briefly heated in a furnace without the presence of oxygen. Typically, the reaction temperature is very high, at around 850°C, but the reaction is only allowed to take place very briefly. In modern cracking furnaces, the residence time is reduced to milliseconds to improve yield, resulting in gas velocities up to the speed of sound. After the cracking temperature has been reached, the gas is quickly quenched to stop the reaction in a transfer line heat exchanger or inside a quenching header using quench oil.

The products produced in the reaction depend on the composition of the feed, the hydrocarbon to steam ratio and on the cracking temperature and furnace residence time. Light hydrocarbon feeds such as ethane, LPGs or light naphtha give product streams rich in the lighter alkenes, including ethylene, propylene, and butadiene. Heavier hydrocarbon (full range and heavy naphthas as well as other refinery products) feeds give some of these, but also give products rich in aromatic hydrocarbons and hydrocarbons suitable for inclusion in gasoline or fuel oil.

A higher cracking temperature (also referred to as severity) favors the production of ethene and benzene, whereas lower severity produces higher amounts of propene, C4-hydrocarbons and liquid products. The process also results in the slow deposition of coke, a form of carbon, on the reactor walls. This degrades the efficiency of the reactor, so reaction conditions are designed to minimize this. Nonetheless, a steam cracking furnace can usually only run for a few months at a time between de-cokings. Decokes require the furnace to be isolated from the process and then a flow of steam or a steam/air mixture is passed through the furnace coils. This converts the hard solid carbon layer to carbon monoxide and carbon dioxide. Once this reaction is complete, the furnace can be returned to service.

Catalytic Methods

The catalytic cracking process involves the presence of acidcatalysts (usually solid acids such as silica-alumina and zeolites) which promote a heterolytic (asymmetric) breakage of bonds yielding pairs of ions of opposite charges, usually a carbocation and the very unstable hydrideanion. Carbon-localized free radicals and cations are both highly unstable and undergo processes of chain rearrangement, C-C scission in position beta as in cracking, and intra- and intermolecular hydrogen transfer. In both types of processes, the corresponding reactive intermediates (radicals, ions) are permanently regenerated, and thus they proceed by a self-propagating chain mechanism. The chain of reactions is eventually terminated by radical or ion recombination.

Fluid Catalytic Cracking

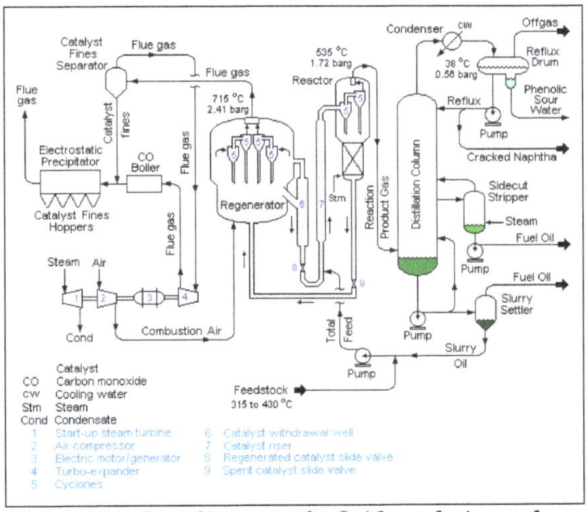

Schematic flow diagram of a fluid catalytic cracker.

Fluid catalytic cracking is a commonly used process, and a modern oil refinery will typically include a *cat cracker*, particularly at refineries in the US, due to the high demand for gasoline. The process was first used around 1942 and employs a powdered catalyst. During WWII, the Allied Forces had plentiful supplies of the materials in contrast to the Axis Forces which suffered severe shortages of gasoline and artificial rubber. Initial process implementations were based on low activity alumina catalyst and a reactor where the catalyst particles were suspended in a rising flow of feed hydrocarbons in a fluidized bed.

Alumina-catalyzed cracking systems are still in use in high school and universitylaboratories in experiments concerning alkanes and alkenes. The catalyst is usually obtained by crushing pumice stones, which contain mainly aluminium oxide and silica into small, porous pieces. In the laboratory, aluminium oxide (or porous pot) must be heated.

In newer designs, cracking takes place using a very active zeolite-based catalyst in a short-contact time vertical or upward-sloped pipe called the "riser". Pre-heated feed is sprayed into the base of the riser via feed nozzles where it contacts extremely hot fluidized catalyst at 1,230 to 1,400 °F (666 to 760 °C). The hot catalyst vaporizes the feed and catalyzes the cracking reactions that break down the high-molecular weight oil into lighter components including LPG, gasoline, and diesel. The catalyst-hydrocarbon mixture flows upward through the riser for a few seconds, and then the mixture is separated via cyclones. The catalyst-free hydrocarbons are routed to a main fractionator for separation into fuel gas, LPG, gasoline, naphtha, light cycle oils used in diesel and jet fuel, and heavy fuel oil.

During the trip up the riser, the cracking catalyst is "spent" by reactions which deposit coke on the catalyst and greatly reduce activity and selectivity. The "spent" catalyst is disengaged from the cracked hydrocarbon vapors and sent to a stripper where it contacts steam to remove hydrocarbons remaining in the catalyst pores. The "spent" catalyst then flows into a fluidized-bed regenerator where air (or in some cases air plus oxygen) is used to burn off the coke to restore catalyst activity and also provide the necessary heat for the next reaction cycle, cracking being an endothermic reaction. The "regenerated" catalyst then flows to the base of the riser, repeating the cycle.

The gasoline produced in the FCC unit has an elevated octane rating but is less chemically stable compared to other gasoline components due to its olefinic profile. Olefins in gasoline are responsible for the formation of polymeric deposits in storage tanks, fuel ducts and injectors. The FCC LPG is an important source of C_3-C_4 olefins and isobutane that are essential feeds for the alkylation process and the production of polymers such as polypropylene.

Hydrocracking

Hydrocracking is a catalytic cracking process assisted by the presence of added hydrogen gas. Unlike a hydrotreater, where hydrogen is used to cleave C-S and C-N bonds, hydrocracking uses hydrogen to break C-C bonds (hydrotreatment is conducted prior to hydrocracking to protect the catalysts in a hydrocracking process).

The products of this process are saturated hydrocarbons; depending on the reaction conditions (temperature, pressure, catalyst activity) these products range from ethane, LPG to heavier hydrocarbons consisting mostly of isoparaffins. Hydrocracking is normally facilitated by a bifunctional catalyst that is capable of rearranging and breaking hydrocarbon chains as well as adding hydrogen to aromatics and olefins to produce naphthenes and alkanes.

The major products from hydrocracking are jet fuel and diesel, but low sulphur naphtha fractions and LPG are also produced. All these products have a very low content of sulfur and other contaminants.

It is very common in Europe and Asia because those regions have high demand for diesel and kerosene. In the US, fluid catalytic cracking is more common because the demand for gasoline is higher.

The hydrocracking process depends on the nature of the feedstock and the relative rates of the two competing reactions, hydrogenation and cracking. Heavy aromatic feedstock is converted into lighter products under a wide range of very high pressures (1,000-2,000 psi) and fairly high temperatures (750°-1,500° F), in the presence of hydrogen and special catalysts.

The primary function of hydrogen is, thus: a) preventing the formation of polycyclic aromatic compounds if feedstock has a high paraffinic content. b) reduced tar formation c) reducing impurities d) preventing buildup of coke on the catalyst. e) converting sulfur and nitrogen compounds present in the feedstock to hydrogen sulfide and ammonia, and e) achieving high cetane number fuel.

Fluid Catalytic Cracking

A typical fluid catalytic cracking unit in a petroleum refinery.

Fluid catalytic cracking (FCC) is one of the most important conversion processes used in petroleum refineries. It is widely used to convert the high-boiling, high-molecular weight hydrocarbon fractions of petroleumcrude oils to more valuable gasoline, olefinic gases, and other products. Cracking of petroleum hydrocarbons was originally done by thermal cracking, which has been almost completely replaced by catalytic cracking because it produces more gasoline with a higher octane rating. It also produces byproduct gases that are more olefinic, and hence more valuable, than those produced by thermal cracking.

The feedstock to an FCC is usually that portion of the crude oil that has an initial boiling point of 340 °C or higher at atmospheric pressure and an average molecular weight ranging from about 200 to 600 or higher. This portion of crude oil is often referred to as heavy gas oil or vacuum gas oil (HVGO). The FCC process vaporizes and breaks the long-chain molecules of the high-boiling hydrocarbon liquids into much shorter molecules by contacting the feedstock, at high temperature and moderate pressure, with a fluidized powdered catalyst.

Economics

Oil refineries use fluid catalytic cracking to correct the imbalance between the market demand for gasoline and the excess of heavy, high boiling range products resulting from the distillation of crude oil.

As of 2006, FCC units were in operation at 400 petroleum refineries worldwide and about one-third of the crude oil refined in those refineries is processed in an FCC to produce high-octane gasoline and fuel oils. During 2007, the FCC units in the United States processed a total of 5,300,000 barrels (840,000 m^3) per day of feedstock and FCC units worldwide processed about twice that amount.

FCC units are less common in Europe and Asia because those regions have high demand for diesel and kerosene, which can be satisfied with hydrocracking. In the US, fluid catalytic cracking is more common because the demand for gasoline is higher.

Flow Diagram and Process Description

The modern FCC units are all continuous processes which operate 24 hours a day for as long as 3 to 5 years between scheduled shutdowns for routine maintenance.

There are several different proprietary designs that have been developed for modern FCC units. Each design is available under a license that must be purchased from the design developer by any petroleum refining company desiring to construct and operate an FCC of a given design.

There are two different configurations for an FCC unit: the "stacked" type where the reactor and the catalyst regenerator are contained in a single vessel with the reactor above the catalyst regenerator and the "side-by-side" type where the reactor and catalyst regenerator are in two separate vessels. These are the major FCC designers and licensors:

Side-by-side configuration:

- CB&I.
- ExxonMobil Research and Engineering (EMRE).
- Shell Global Solutions.
- Axens / Stone & Webster Process Technology — currently owned by Technip.
- Universal Oil Products (UOP) — currently fully owned subsidiary of Honeywell.

Stacked configuration:

- Kellogg Brown & Root (KBR).

Each of the proprietary design licensors claims to have unique features and advantages. A complete discussion of the relative advantages of each of the processes is beyond the scope of this article.

Reactor and Regenerator

The reactor and regenerator are considered to be the heart of the fluid catalytic cracking unit. The schematic flow diagram of a typical modern FCC unit in Figure 1 below is based upon the "side-by-side" configuration. The preheated high-boiling petroleum feedstock (at about 315 to 430 °C) consisting of long-chain hydrocarbon molecules is combined with recycle slurry oil from the bottom of the distillation column and injected into the *catalyst riser* where it is vaporized and cracked into smaller molecules of vapor by contact and mixing with the very hot powdered catalyst from the regenerator. All of the cracking reactions take place in the catalyst riser within a period of 2–4 seconds. The hydrocarbon vapors "fluidize" the powdered catalyst and the mixture of hydrocarbon vapors and catalyst flows upward to enter the *reactor* at a temperature of about 535 °C and a pressure of about 1.72 bar.

The reactor is a vessel in which the cracked product vapors are: (a) separated from the so-called *spent catalyst* by flowing through a set of two-stage cyclones within the reactor and (b) the *spent catalyst* flows downward through a steam stripping section to remove any hydrocarbon vapors before the spent catalyst returns to the *catalyst regenerator*. The flow of spent catalyst to the regenerator is regulated by a slide valve in the spent catalyst line.

Since the cracking reactions produce some carbonaceous material (referred to as catalyst coke) that deposits on the catalyst and very quickly reduces the catalyst reactivity, the catalyst is regenerated by burning off the deposited coke with air blown into the regenerator. The regenerator operates at a temperature of about 715 °C and a pressure of about 2.41 barg, hence the regenerator operates at about 0.7 barg higher pressure than the reactor. The combustion of the coke is exothermic and it produces a large amount of heat that is partially absorbed by the regenerated catalyst and provides the heat required for the vaporization of the feedstock and the endothermic cracking reactions that take place in the catalyst riser. For that reason, FCC units are often referred to as being 'heat balanced'.

The hot catalyst (at about 715 °C) leaving the regenerator flows into a *catalyst withdrawal well* where any entrained combustion flue gases are allowed to escape and flow back into the upper part to the regenerator. The flow of regenerated catalyst to the feedstock injection point below the catalyst riser is regulated by a slide valve in the regenerated catalyst line. The hot flue gas exits the regenerator after passing through multiple sets of two-stage cyclones that remove entrained catalyst from the flue gas.

The amount of catalyst circulating between the regenerator and the reactor amounts to about 5 kg per kg of feedstock, which is equivalent to about 4.66 kg per litre of feedstock. Thus, an FCC unit processing 75,000 barrels per day (11,900 m^3/d) will circulate about 55,900 tonnes per day of catalyst.

Distillation Column

The reaction product vapors (at 535 °C and a pressure of 1.72 barg) flow from the top of the reactor to the bottom section of the distillation column (commonly referred to as the *main fractionator*) where they are distilled into the FCC end products of cracked naphtha, fuel oil, and offgas. After further processing for removal of sulfur compounds, the cracked naphtha becomes a high-octane component of the refinery's blended gasolines.

The main fractionator offgas is sent to what is called a *gas recovery unit* where it is separated into butanes and butylenes, propane and propylene, and lower molecular weight gases (hydrogen, methane, ethylene and ethane). Some FCC gas recovery units may also separate out some of the ethane and ethylene.

Although the schematic flow diagram above depicts the main fractionator as having only one sidecut stripper and one fuel oil product, many FCC main fractionators have two sidecut strippers and produce a light fuel oil and a heavy fuel oil. Likewise, many FCC main fractionators produce a light cracked naphtha and a heavy cracked naphtha. The terminology *light* and *heavy* in this context refers to the product boiling ranges, with light products having a lower boiling range than heavy products.

The bottom product oil from the main fractionator contains residual catalyst particles which were not completely removed by the cyclones in the top of the reactor. For that reason, the bottom product oil is referred to as a *slurry oil*. Part of that slurry oil is recycled back into the main fractionator above the entry point of the hot reaction product vapors so as to cool and partially condense the reaction product vapors as they enter the main fractionator. The remainder of the slurry oil is pumped through a slurry settler. The bottom oil from the slurry settler contains most of the slurry oil catalyst particles and is recycled back into the catalyst riser by combining it with the FCC feedstock oil. The so-called *clarified slurry oil* or decant oil is withdrawn from the top of slurry settler for use elsewhere in the refinery, as a heavy fuel oil blending component, or as carbon black feedstock.

Regenerator Flue Gas

Depending on the choice of FCC design, the combustion in the regenerator of the coke on the spent catalyst may or may not be complete combustion to carbon dioxide CO_2. The combustion air flow is controlled so as to provide the desired ratio of carbon monoxide (CO) to carbon dioxide for each specific FCC design.

In the design shown in Figure 1, the coke has only been partially combusted to CO_2. The combustion flue gas (containing CO and CO_2) at 715 °C and at a pressure of 2.41 barg is routed through a secondary catalyst separator containing *swirl tubes* designed to remove 70 to 90 percent of the particulates in the flue gas leaving the regenerator. This is required to prevent erosion damage to the blades in the turbo-expander that the flue gas is next routed through.

The expansion of flue gas through a turbo-expander provides sufficient power to drive the regenerator's combustion air compressor. The electrical motor-generator can consume or produce electrical power. If the expansion of the flue gas does not provide enough power to drive the air compressor, the electric motor/generator provides the needed additional power. If the flue gas expansion provides more power than needed to drive the air compressor, than the electric motor/generator converts the excess power into electric power and exports it to the refinery's electrical system.

The expanded flue gas is then routed through a steam-generating boiler (referred to as a *CO boiler*) where the carbon monoxide in the flue gas is burned as fuel to provide steam for use in the refinery as well as to comply with any applicable environmental regulatory limits on carbon monoxide emissions.

The flue gas is finally processed through an electrostatic precipitator (ESP) to remove residual particulate matter to comply with any applicable environmental regulations regarding particulate emissions. The ESP removes particulates in the size range of 2 to 20 μm from the flue gas. Particulate filter systems, known as Fourth Stage Separators (FSS) are sometimes required to meet particulate emission limits. These can replace the ESP when particulate emissions are the only concern.

The steam turbine in the flue gas processing system (shown in the above diagram) is used to drive the regenerator's combustion air compressor during start-ups of the FCC unit until there is sufficient combustion flue gas to take over that task.

Chemistry

Before delving into the chemistry involved in catalytic cracking, it will be helpful to briefly discuss the composition of petroleum crude oil.

Petroleum crude oil consists primarily of a mixture of hydrocarbons with small amounts of other organic compounds containing sulfur, nitrogen and oxygen. The crude oil also contains small amounts of metals such as copper, iron, nickel and vanadium.

Table 1	
Carbon	83-87%
Hydrogen	10-14%
Nitrogen	0.1-2%
Oxygen	0.1-1.5%
Sulfur	0.5-6%
Metals	< 0.1%

The elemental composition ranges of crude oil are summarized in Table 1 and the hydrocarbons in the crude oil can be classified into three types:

- Paraffins or alkanes: saturated straight-chain or branched hydrocarbons, without any ring structures.
- Naphthenes or cycloalkanes: saturated hydrocarbons having one or more ring structures with one or more side-chain paraffins.
- Aromatics: hydrocarbons having one or more unsaturated ring structures such as benzene or unsaturated polycyclic ring structures such as naphthalene or phenanthrene, any of which may also have one or more side-chain paraffins.

Olefins or alkenes, which are unsaturated straight-chain or branched hydrocarbons, do not occur naturally in crude oil.

In plain language, the fluid catalytic cracking process breaks large hydrocarbon molecules into smaller molecules by contacting them with powdered catalyst at a high temperature and moderate pressure which first vaporizes the hydrocarbons and then breaks them. The cracking reactions occur in the vapor phase and start immediately when the feedstock is vaporized in the catalyst riser.

Figure 2 is a very simplified schematic diagram that exemplifies how the process breaks high boiling, straight-chain alkane (paraffin) hydrocarbons into smaller straight-chain alkanes as well as branched-chain alkanes, branched alkenes (olefins) and cycloalkanes (naphthenes). The breaking of the large hydrocarbon molecules into smaller molecules is more technically referred to by organic chemists as *scission* of the carbon-to-carbon bonds.

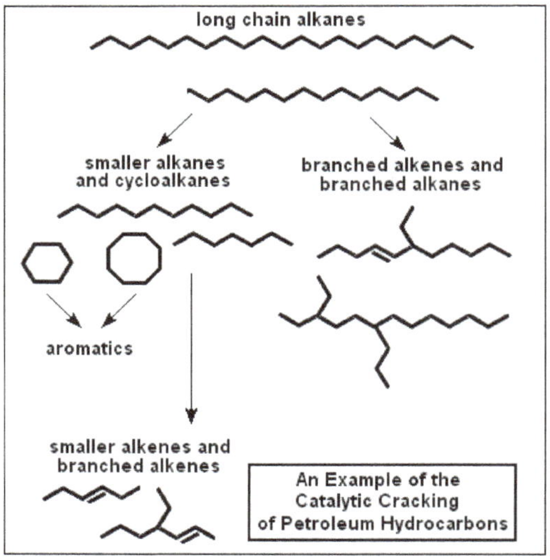

Figure: Diagrammatic example of the catalytic cracking of petroleum hydrocarbons.

As depicted in Figure, some of the smaller alkanes are then broken and converted into even smaller alkenes and branched alkenes such as the gases ethylene, propylene, butylenes, and isobutylenes. Those olefinic gases are valuable for use as petrochemical feedstocks. The propylene, butylene and isobutylene are also valuable feedstocks for certain petroleum refining processes that convert them into high-octane gasoline blending components.

As also depicted in Figure, the cycloalkanes (naphthenes) formed by the initial breakup of the large molecules are further converted to aromatics such as benzene, toluene, and xylenes, which boil in the gasoline boiling range and have much higher octane ratings than alkanes.

In the cracking process carbon is also produced which gets deposited on the catalyst (catalyst coke). The carbon formation tendency or amount of carbon in a crude or FCC feed is measured with methods such as Micro Carbon Residue, Conradson Carbon Residue or Ramsbottom Carbon Residue.

By no means does Figure include all the chemistry of the primary and secondary reactions taking place in the fluid catalytic process. There are a great many other reactions involved. However, a full discussion of the highly technical details of the various catalytic cracking reactions is beyond the scope of this article and can be found in the technical literature.

Catalysts

Modern FCC catalysts are fine powders with a bulk density of 0.80 to 0.96 g/cm^3 and having a particle size distribution ranging from 10 to 150 μm and an average particle size of 60 to 100 μm. The design and operation of an FCC unit is largely dependent upon the chemical and physical properties of the catalyst. The desirable properties of an FCC catalyst are:

- Good stability to high temperature and to steam.

- High activity.
- Large pore sizes.
- Good resistance to attrition.
- Low coke production.

A modern FCC catalyst has four major components: crystalline zeolite, matrix, binder, and filler. Zeolite is the primary active component and can range from about 15 to 50 weight percent of the catalyst. The zeolite used in FCC catalysts is referred to as faujasite or as *Type Y* and is composed of silica and alumina tetrahedra with each tetrahedron having either an aluminum or a silicon atom at the center and four oxygen atoms at the corners. It is a molecular sieve with a distinctive lattice structure that allows only a certain size range of hydrocarbon molecules to enter the lattice. In general, the zeolite does not allow molecules larger than 8 to 10 nm (i.e., 80 to 100 ångströms) to enter the lattice.

The catalytic sites in the zeolite are strong acids (equivalent to 90% sulfuric acid) and provide most of the catalytic activity. The acidic sites are provided by the alumina tetrahedra. The aluminum atom at the center of each alumina tetrahedra is at a +3 oxidation state surrounded by four oxygen atoms at the corners which are shared by the neighboring tetrahedra. Thus, the net charge of the alumina tetrahedra is -1 which is balanced by a sodiumion during the production of the catalyst. The sodium ion is later replaced by an ammonium ion, which is vaporized when the catalyst is subsequently dried, resulting in the formation of Lewis and Brønsted acidic sites. In some FCC catalysts, the Brønsted sites may be later replaced by rare earth metals such as cerium and lanthanum to provide alternative activity and stability levels. The acidic activity of the catalyst is deactivated by the nitrogen present in the feedstock. Typically the nitrogen content of the feedstock is between 1000 and 2000 ppm. Higher Nitrogen content results in significant depression of the conversion rate.

The matrix component of an FCC catalyst contains amorphous alumina which also provides catalytic activity sites and in larger pores that allows entry for larger molecules than does the zeolite. That enables the cracking of higher-boiling, larger feedstock molecules than are cracked by the zeolite.

The binder and filler components provide the physical strength and integrity of the catalyst. The binder is usually silica sol and the filler is usually a clay (kaolin).

Nickel, vanadium, iron, copper and other metal contaminants, present in FCC feedstocks in the parts per million range, all have detrimental effects on the catalyst activity and performance. Nickel and vanadium are particularly troublesome. There are a number of methods for mitigating the effects of the contaminant metals:

- Avoid feedstocks with high metals content: This seriously hampers a refinery's flexibility to process various crude oils or purchased FCC feedstocks.
- Feedstock feed pretreatment: Hydrodesulfurization of the FCC feedstock removes some of the metals and also reduces the sulfur content of the FCC products. However, this is quite a costly option.

- Increasing fresh catalyst addition: All FCC units withdraw some of the circulating *equilibrium catalyst* as spent catalyst and replaces it with fresh catalyst in order to maintain a desired level of activity. Increasing the rate of such exchange lowers the level of metals in the circulating *equilibrium catalyst*, but this is also quite a costly option.

- Demetallization: The commercial proprietary *Demet Process* removes nickel and vanadium from the withdrawn spent catalyst. The nickel and vanadium are converted to chlorides which are then washed out of the catalyst. After drying, the demetallized catalyst is recycled into the circulating catalyst. Removals of about 95 percent nickel removal and 67 to 85 percent vanadium have been reported. Despite that, the use of the Demet process has not become widespread, perhaps because of the high capital expenditure required.

- Metals passivation: Certain materials can be used as additives which can be impregnated into the catalyst or added to the FCC feedstock in the form of metal-organic compounds. Such materials react with the metal contaminants and passivate the contaminants by forming less harmful compounds that remain on the catalyst. For example, antimony and bismuth are effective in passivating nickel and tin is effective in passivating vanadium. A number of proprietary passivation processes are available and fairly widely used.

The major suppliers of FCC catalysts worldwide include Albemarle Corporation, W.R. Grace Company and BASF Catalysts (formerly Engelhard). The price for lanthanum oxide used in fluid catalytic cracking has risen from $5 per kilogram in early 2010 to $140 per kilogram in June 2011.

History

The first commercial use of catalytic cracking occurred in 1915 when Almer M. McAfee of Gulf Refining Company developed a batch process using aluminum chloride (a Friedel Crafts catalyst known since 1877) to catalytically crack heavy petroleum oils. However, the prohibitive cost of the catalyst prevented the widespread use of McAfee's process at that time.

In 1922, a French mechanical engineer named Eugene Jules Houdry and a French pharmacist named E.A. Prudhomme set up a laboratory near Paris to develop a catalytic process for converting lignite coal to gasoline. Supported by the French government, they built a small demonstration plant in 1929 that processed about 60 tons per day of lignite coal. The results indicated that the process was not economically viable and it was subsequently shut down.

Houdry had found that Fuller's earth, a clay mineral containing aluminosilicates, could convert oil derived from the lignite to gasoline. He then began to study the catalysis of petroleum oils and had some success in converting vaporized petroleum oil to gasoline. In 1930, the Vacuum Oil Company invited him to come to the United States and he moved his laboratory to Paulsboro, New Jersey.

In 1931, the Vacuum Oil Company merged with Standard Oil of New York (Socony) to form the Socony-Vacuum Oil Company. In 1933, a small Houdry unit processed 200 barrels per day (32 m³/d) of petroleum oil. Because of the economic depression of the early 1930s, Socony-Vacuum was no longer able to support Houdry's work and gave him permission to seek help elsewhere.

In 1933, Houdry and Socony-Vacuum joined with Sun Oil Company in developing the Houdry process. Three years later, in 1936, Socony-Vacuum converted an older thermal cracking unit in their Paulsboro refinery in New Jersey to a small demonstration unit using the Houdry process to catalytically crack 2,000 barrels per day (320 m^3/d) of petroleum oil.

In 1937, Sun Oil began operation of a new Houdry unit processing 12,000 barrels per day (1,900 m^3/d) in their Marcus Hook refinery in Pennsylvania. The Houdry process at that time used reactors with a fixed bed of catalyst and was a semi-batch operation involving multiple reactors with some of the reactors in operation while other reactors were in various stages of regenerating the catalyst. Motor-driven valves were used to switch the reactors between online operation and offline regeneration and a cycle timer managed the switching. Almost 50 percent of the cracked product was gasoline as compared with about 25 percent from the thermal cracking processes.

By 1938, when the Houdry process was publicly announced, Socony-Vacuum had eight additional units under construction. Licensing the process to other companies also began and by 1940 there were 14 Houdry units in operation processing 140,000 barrels per day (22,000 m^3/d).

The next major step was to develop a continuous process rather than the semi-batch Houdry process. That step was implemented by advent of the moving-bed process known as the Thermofor Catalytic Cracking (TCC) process which used a bucket conveyor-elevator to move the catalyst from the regeneration kiln to the separate reactor section. A small semicommercial demonstration TCC unit was built in Socony-Vacuum's Paulsboro refinery in 1941 and operated successfully, producing 500 barrels per day (79 m^3/d). Then a full-scale commercial TCC unit processing 10,000 barrels per day (1,600 m^3/d) began operation in 1943 at the Beaumont, Texas refinery of Magnolia Oil Company, an affiliate of Socony-Vacuum. By the end of World War II in 1945, the processing capacity of the TCC units in operation was about 300,000 barrels per day (48,000 m^3/d).

It is said that the Houdry and TCC units were a major factor in the winning of World War II by supplying the high-octane gasoline needed by the air forces of Great Britain and the United States for the more efficient higher compression ratio engines of the Spitfire and the Mustang.

In the years immediately after World War II, the Houdriflow process and the air-lift TCC process were developed as improved variations on the moving-bed theme. Just like Houdry's fixed-bed reactors, the moving-bed designs were prime examples of good engineering by developing a method of continuously moving the catalyst between the reactor and regeneration sections. The first air-lift TCC unit began operation in October 1950 at the Beaumont, Texas refinery.

This fluid catalytic cracking process had first been investigated in the 1920s by Standard Oil of New Jersey, but research on it was abandoned during the economic depression years of 1929 to 1939. In 1938, when the success of Houdry's process had become apparent, Standard Oil of New Jersey resumed the project as part of a consortium of that include five oil companies (Standard Oil of New Jersey, Standard Oil of Indiana, Anglo-Iranian Oil, Texas Oil and Dutch Shell), two engineering-construction companies (M.W. Kellogg and Universal Oil Products) and a German chemical company (I.G. Farben). The consortium was called Catalytic Research Associates (CRA) and its purpose was to develop a catalytic cracking process which would not impinge on Houdry's patents.

Chemical engineering professors Warren K. Lewis and Edwin R. Gilliland of the Massachusetts Institute of Technology (MIT) suggested to the CRA researchers that a low velocity gas flow through a powder might "lift" it enough to cause it to flow in a manner similar to a liquid. Focused on that idea of a fluidized catalyst, researchers Donald Campbell, Homer Martin, Eger Murphree and

Charles Tyson of the Standard Oil of New Jersey (now Exxon-Mobil Company) developed the first fluidized catalytic cracking unit. Their U.S. Patent No. 2,451,804, *A Method of and Apparatus for Contacting Solids and Gases*, describes their milestone invention. Based on their work, M. W. Kellogg Company constructed a large pilot plant in the Baton Rouge, Louisiana refinery of the Standard Oil of New Jersey. The pilot plant began operation in May 1940.

Based on the success of the pilot plant, the first commercial fluid catalytic cracking plant (known as the Model I FCC) began processing 13,000 barrels per day (2,100 m^3/d) of petroleum oil in the Baton Rouge refinery on May 25, 1942, just four years after the CRA consortium was formed and in the midst of World War II. A little more than a month later, in July 1942, it was processing 17,000 barrels per day (2,700 m^3/d). In 1963, that first Model I FCC unit was shut down after 21 years of operation and subsequently dismantled.

In the many decades since the Model I FCC unit began operation, the fixed bed Houdry units have all been shut down as have most of the moving bed units (such as the TCC units) while hundreds of FCC units have been built. During those decades, many improved FCC designs have evolved and cracking catalysts have been greatly improved, but the modern FCC units are essentially the same as that first Model I FCC unit.

Note: All of the refinery and company names in this history section (with the exception of Universal Oil Products) have changed over time by mergers and buyouts. Some have changed a number of times.

Gas Flare

Flare stack at the Shell Haven refinery in England.

A gas flare, alternatively known as a flare stack, is a gas combustion device used in industrial plants such as petroleum refineries, chemical plants, natural gas processing plants as well as at oil or gas production sites having oil wells, gas wells, offshore oil and gas rigs and landfills.

North Dakota Flaring of Gas.

In industrial plants, flare stacks are primarily used for burning off flammable gas released by pressure relief valves during unplanned over-pressuring of plant equipment. During plant or partial plant startups and shutdowns, flare stacks are also often used for the planned combustion of gases over relatively short periods.

A great deal of gas flaring at many oil and gas production sites has to do with protection against the dangers of over-pressuring industrial plant equipment. When petroleum crude oil is extracted and produced from onshore or offshore oil wells, raw natural gas associated with the oil is brought to the surface as well. Especially in areas of the world lacking pipelines and other gas transportation infrastructure, vast amounts of such associated gas are commonly flared as waste or unusable gas. The flaring of associated gas may occur at the top of a vertical flare stack (as in the adjacent photo) or it may occur in a ground-level flare in an earthen pit. Preferably, associated gas is reinjected into the reservoir, which saves it for future use while maintaining higher well pressure and crude oil producibility.

Overall Flare System in Industrial Plants

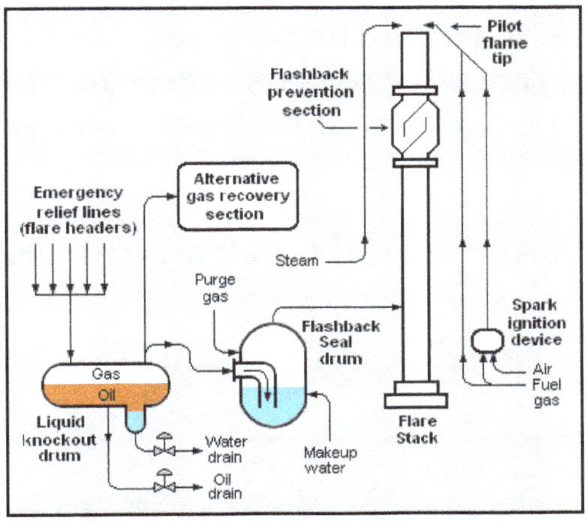

Schematic flow diagram of an overall vertical, elevated flare stack system in an industrial plant.

When industrial plant equipment items are over-pressured, the pressure relief valve is an essential safety device that automatically release gases and sometimes liquids. Those pressure relief valves are required by industrial design codes and standards as well as by law.

The released gases and liquids are routed through large piping systems called *flare headers* to a vertical elevated flare. The released gases are burned as they exit the flare stacks. The size and brightness of the resulting flame depends upon the flammable material's flow rate in joules per hour (or btu per hour).

Most industrial plant flares have a vapor-liquid separator (also known as a knockout drum) upstream of the flare to remove any large amounts of liquid that may accompany the relieved gases.

Steam is very often injected into the flame to reduce the formation of black smoke. When too much steam is added, a condition known as "over steaming" can occur resulting in reduced combustion efficiency and higher emissions. To keep the flare system functional, a small amount of gas is continuously burned, like a pilot light, so that the system is always ready for its primary purpose as an over-pressure safety system.

The adjacent flow diagram depicts the typical components of an overall industrial flare stack system:

- A knockout drum to remove any oil or water from the relieved gases.
- A water seal drum to prevent any flashback of the flame from the top of the flare stack.
- An alternative gas recovery system for use during partial plant startups and shutdowns as well as other times when required. The recovered gas is routed into the fuel gas system of the overall industrial plant.
- A steam injection system to provide an external momentum force used for efficient mixing of air with the relieved gas, which promotes smokeless burning.
- A pilot flame (with its ignition system) that burns all the time so that it is available to ignite relieved gases when needed.
- The flare stack, including a flashback prevention section at the upper part of the stack.

There is also a safe method to divert the flare gas which is insertion of Liquid U seal with Liquid Hold up vessel. The Liquid U seal is designed to take pressure up to permitted back pressure of the system. This helps to divert the flare gas to recovery system. In case of plant upset, pressure rises and liquid in the U seal will move into Liquid Hold up vessel. On normalization, the Liquid U seal will start diverting the gas again.

Impacts of Waste Flaring Associated Gas from Oil Drilling Sites and Other Facilities

Flaring of associated gas from an oil well site in Nigeria.

Flaring gases from an oil platform in the North Sea.

Improperly operated flares may emit methane and other volatile organic compounds as well as sulfur dioxide and other sulfur compounds, which are known to exacerbate asthma and other respiratory problems. Other emissions from improperly operated flares may include, aromatic hydrocarbons (benzene, toluene, xylenes) and benzapyrene, which are known to be carcinogenic.

Flaring can affect wildlife by attracting birds and insects to the flame. Approximately 7,500 migrating songbirds were attracted to and killed by the flare at the liquefied natural gas terminal in Saint John, New Brunswick, Canada on September 13, 2013. Similar incidents have occurred at flares on offshore oil and gas installations. Moths are known to be attracted to lights. A brochure published by the Secretariat of the Convention on Biological Diversity describing the Global Taxonomy Initiative describes a situation where "*a taxonomist working in a tropical forest noticed that a gas flare at an oil refinery was attracting and killing hundreds of these [hawk or sphinx] moths. Over the course of the months and years that the refinery was running a vast number of moths must have been killed, suggesting that plants could not be pollinated over a large area of forest*".

As of the end of 2011, 150 × 10^9 cubic meters (5.3 × 10^{12} cubic feet) of associated gas are flared annually. That is equivalent to about 25 per cent of the annual natural gas consumption in the United States or about 30 per cent of the annual gas consumption in the European Union. If it were to reach market, this quantity of gas (at a nominal value of $5.62 per 1000 cubic feet) would be worth $29.8 billion USD.

Also as of the end of 2011, 10 countries accounted for 72 per cent of the flaring, and twenty for 86 per cent. The top ten leading contributors to world gas flaring at the end of 2011, were (in declining order): Russia (27%), Nigeria (11%), Iran (8%), Iraq (7%), United States (5%), Algeria (4%), Kazakhstan (3%), Angola (3%), Saudi Arabia (3%) and Venezuela (3%).

That amount of flaring and burning of associated gas from oil drilling sites is a significant source of carbon dioxide (CO_2) emissions. Coupled with fossil fuel combustion and cement production, flaring's carbon dioxide emissions in 2010 have tripled (1300 ± 110 GtCO2) compared to the last recording (years 1750-1970, 420 ± 35 GtCO had been emitted.) 2400 × 10^6 tons of carbon dioxide are emitted annually in this way and it amounts to about 1.2 per cent of the worldwide emissions of carbon dioxide. That may seem to be insignificant, but in perspective it is more than half of the *Certified Emissions Reductions* (a type of carbon credits) that have been issued under the rules and mechanisms of the Kyoto Protocol as of June 2011.

Satellite data show that from 2005 to 2010, global gas flaring decreased by about 20%. The most significant reductions in terms of volume were made in Russia (down 40%) and Nigeria (down 29%).

Hydrodenitrogenation

Hydrodenitrogenation (HDN) is an industrial process for the removal of nitrogen from petroleum. Organonitrogen compounds, even though they occur at low levels, are undesirable because they poisoning with downstream catalysts. Furthermore, upon combustion, organonitrogen compounds generate NOx, a pollutant. HDN is effected as general hydroprocessing, which traditionally focuses on hydrodesulfurization (HDS) because sulfur compounds are even more problematic. To some extent, hydrodeoxygenation (HDO) is also effected.

Typical organonitrogen compounds in petroleum include quinolines and porphyrins and their derivatives. The total nitrogen content is typically less than 1% and the targeted levels are in the ppm range. As described in organic geochemistry, organonitrogen compounds are derivatives or degradation products of the compounds in the living matter that comprised the precursor to fossil fuels. In HDN, the organonitrogen compounds are treated at high temperatures with hydrogen in the presence of a catalyst, the net transformation being:

$$R_3N + 3\,H_2 \rightarrow 3\,RH + NH_3$$

The catalysts generally consist of cobalt and nickel as well as molybdenum disulfide or less often tungsten disulfidesupported on alumina]]. The precise composition of the catalyst, i.e. Co/Ni and Mo/W ratios, are tuned for particular feedstocks. A wide variety of catalyst compositions have been considered, including metal phosphides.

Hydrodesulfurization

Hydrodesulfurization (HDS) is a catalytic chemical process widely used to remove sulfur (S) from natural gas and from refined petroleum products, such as gasoline or petrol, jet fuel, kerosene, diesel fuel, and fuel oils. The purpose of removing the sulfur is to reduce the sulfur dioxide (SO 2) emissions that result from using those fuels in automotive vehicles, aircraft, railroad locomotives, ships, gas or oil burning power plants, residential and industrial furnaces, and other forms of fuel combustion.

Another important reason for removing sulfur from the naphtha streams within a petroleum refinery is that sulfur, even in extremely low concentrations, poisons the noble metal catalysts (platinum and rhenium) in the catalytic reforming units that are subsequently used to upgrade the octane rating of the naphtha streams.

The industrial hydrodesulfurization processes include facilities for the capture and removal of the resulting hydrogen sulfide (H2S) gas. In petroleum refineries, the hydrogen sulfide gas is then subsequently converted into byproduct elemental sulfur or sulfuric acid (H2SO4). In fact, the vast majority of the 64,000,000 metric tons of sulfur produced worldwide in 2005 was byproduct sulfur from refineries and other hydrocarbon processing plants.

An HDS unit in the petroleum refining industry is also often referred to as a hydrotreater.

History

Although some reactions involving catalytic hydrogenation of organic substances were already known, the property of finely divided nickel to catalyze the fixation of hydrogen on hydrocarbon (ethylene, benzene) double bonds was discovered by the French chemistPaul Sabatier in 1897. Through this work, he found that unsaturated hydrocarbons in the vapor phase could be converted into saturated hydrocarbons by using hydrogen and a catalytic metal, laying the foundation of the modern catalytic hydrogenation process.

Soon after Sabatier's work, a German chemist, Wilhelm Normann, found that catalytic hydrogenation could be used to convert unsaturated fatty acids or glycerides in the liquid phase into saturated ones. He was awarded a patent in Germany in 1902 and in Britain in 1903, which was the beginning of what is now a worldwide industry.

In the mid-1950s, the first noble metal catalytic reforming process (the Platformer process) was commercialized. At the same time, the catalytic hydrodesulfurization of the naphtha feed to such reformers was also commercialized. In the decades that followed, various proprietary catalytic hydrodesulfurization processes, such as the one depicted in the flow diagram below, have been commercialized. Currently, virtually all of the petroleum refineries worldwide have one or more HDS units.

By 2006, miniature microfluidic HDS units had been implemented for treating JP-8 jet fuel to produce clean feed stock for a fuel cellhydrogen reformer. By 2007, this had been integrated into an operating 5 kW fuel cell generation system.

Process Chemistry

Hydrogenation is a class of chemical reactions in which the net result is the addition of hydrogen (H). Hydrogenolysis is a type of hydrogenation and results in the cleavage of the C-X chemical bond, where C is a carbon atom and X is a sulfur (S), nitrogen (N) or oxygen (O) atom. The net result of a hydrogenolysis reaction is the formation of C-H and H-X chemical bonds. Thus, hydrodesulfurization is a hydrogenolysis reaction. Using ethanethiol (C2H5SH), a sulfur compound present in some petroleum products, as an example, the hydrodesulfurization reaction can be simply expressed as:

Ethanethiol + Hydrogen	→	Ethane + Hydrogen sulfide
$C_2H_5SH + H_2$	→	$C_2H_6 + H_2S$

For the mechanistic aspects of, and the catalysts used in this reaction see the section catalysts and mechanisms.

Process Description

In an industrial hydrodesulfurization unit, such as in a refinery, the hydrodesulfurization reaction takes place in a fixed-bed reactor at elevated temperatures ranging from 300 to 400 °C and elevated pressures ranging from 30 to 130 atmospheres of absolute pressure, typically in the presence of a catalyst consisting of an alumina base impregnated with cobalt and molybdenum (usually called a CoMo catalyst). Occasionally, a combination of nickel and molybdenum (called NiMo) is used, in addition to the CoMo catalyst, for specific difficult-to-treat feed stocks, such as those containing a high level of chemically bound nitrogen.

The image below is a schematic depiction of the equipment and the process flow streams in a typical refinery HDS unit.

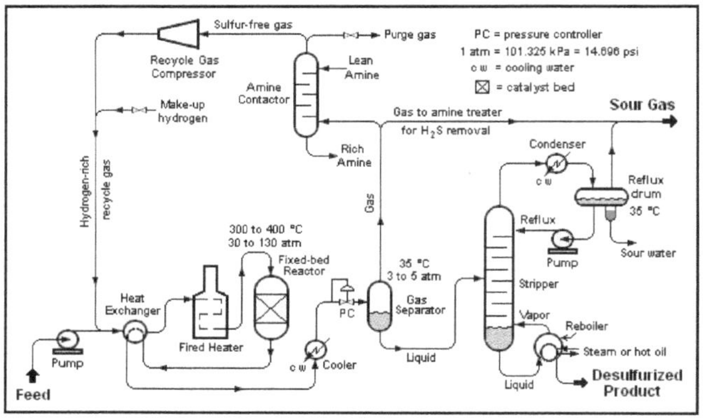

Schematic diagram of a typical Hydrodesulfurization (HDS) unit in a petroleum refinery.

The liquid feed (at the bottom left in the diagram) is pumped up to the required elevated pressure and is joined by a stream of hydrogen-rich recycle gas. The resulting liquid-gas mixture is preheated by flowing through a heat exchanger. The preheated feed then flows through a fired heater where the feed mixture is totally vaporized and heated to the required elevated temperature before entering the reactor and flowing through a fixed-bed of catalyst where the hydrodesulfurization reaction takes place.

The hot reaction products are partially cooled by flowing through the heat exchanger where the reactor feed was preheated and then flows through a water-cooled heat exchanger before it flows through the pressure controller (PC) and undergoes a pressure reduction down to about 3 to 5 atmospheres. The resulting mixture of liquid and gas enters the gas separator vessel at about 35 °C and 3 to 5 atmospheres of absolute pressure.

Most of the hydrogen-rich gas from the gas separator vessel is recycle gas, which is routed through an amine contactor for removal of the reaction product H2S that it contains. The H2S-free hydrogen-rich gas is then recycled back for reuse in the reactor section. Any excess gas from the gas separator vessel joins the sour gas from the stripping of the reaction product liquid.

The liquid from the gas separator vessel is routed through a reboiled stripper distillation tower. The bottoms product from the stripper is the final desulfurized liquid product from hydrodesulfurization unit.

The overhead sour gas from the stripper contains hydrogen, methane, ethane, hydrogen sulfide, propane, and, perhaps, some butane and heavier components. That sour gas is sent to the refinery's central gas processing plant for removal of the hydrogen sulfide in the refinery's main amine gas treating unit and through a series of distillation towers for recovery of propane, butane and pentane or heavier components. The residual hydrogen, methane, ethane, and some propane is used as refinery fuel gas. The hydrogen sulfide removed and recovered by the amine gas treating unit is subsequently converted to elemental sulfur in a Claus process unit or to sulfuric acid in a wet sulfuric acid process or in the conventional Contact Process.

Note that the above description assumes that the HDS unit feed contains no olefins. If the feed does contain olefins (for example, the feed is a naphtha derived from a refinery fluid catalytic cracker (FCC) unit), then the overhead gas from the HDS stripper may also contain some ethene, propene, butenes and pentenes, or heavier components.

It should also be noted that the amine solution to and from the recycle gas contactor comes from and is returned to the refinery's main amine gas treating unit.

Sulfur Compounds in Refinery HDS Feedstocks

The refinery HDS feedstocks (naphtha, kerosene, diesel oil, and heavier oils) contain a wide range of organic sulfur compounds, including thiols, thiophenes, organic sulfides and disulfides, and many others. These organic sulfur compounds are products of the degradation of sulfur containing biological components, present during the natural formation of the fossil fuel, petroleum crude oil.

When the HDS process is used to desulfurize a refinery naphtha, it is necessary to remove the total sulfur down to the parts per million range or lower in order to prevent poisoning the noble metal catalysts in the subsequent catalytic reforming of the naphthas.

When the process is used for desulfurizing diesel oils, the latest environmental regulations in the United States and Europe, requiring what is referred to as ultra-low-sulfur diesel (ULSD), in turn requires that very deep hydrodesulfurization is needed. In the very early 2000s, the governmental regulatory limits for highway vehicle diesel was within the range of 300 to 500 ppm by weight of total sulfur. As of 2006, the total sulfur limit for highway diesel is in the range of 15 to 30 ppm by weight.

Thiophenes

A family of substrates that are particularly common in petroleum are the aromatic sulfur-containing heterocycles called thiophenes. Many kinds of thiophenes occur in petroleum ranging from thiophene itself to more condensed derivatives called benzothiophenes and dibenzothiophenes. Thiophene itself and its alkyl derivatives are easier to hydrogenolyse, whereas dibenzothiophene, especially its 4,6-disubstituted derivatives, are considered the most challenging substrates. Benzothiophenes are midway between the simple thiophenes and dibenzothiophenes in their susceptibility to HDS.

Catalysts and Mechanisms

The main HDS catalysts are based on molybdenum disulfide (MoS2) together with smaller amounts of other metals. The nature of the sites of catalytic activity remains an active area of investigation, but it is generally assumed basal planes of the MoS2 structure are not relevant to catalysis, rather the edges or rims of these sheet. At the edges of the MoS2 crystallites, the molybdenum centre can stabilize a coordinatively unsaturated site (CUS), also known as an anion vacancy. Substrates, such as thiophene, bind to this site and undergo a series of reactions that result in both C-S scission and C=C hydrogenation. Thus, the hydrogen serves multiple roles—generation of anion vacancy by removal of sulfide, hydrogenation, and hydrogenolysis. A simplified diagram for the cycle is shown:

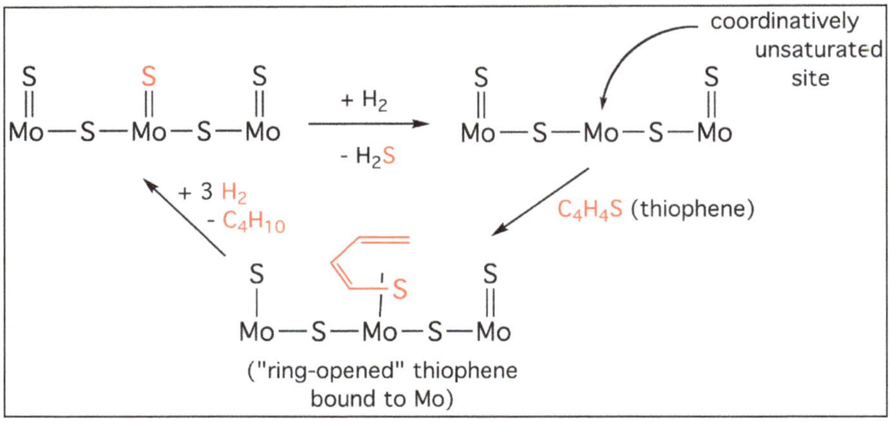

Simplified diagram of a HDS cycle for thiophene.

Catalysts

Most metals catalyse HDS, but it is those at the middle of the transition metal series that are most active. Ruthenium disulfide appears to be the single most active catalyst, but binary combinations of cobalt and molybdenum are also highly active. Aside from the basic cobalt-modified MoS_2 catalyst, nickel and tungsten are also used, depending on the nature of the feed. For example, Ni-W catalysts are more effective for hydrodenitrogenation.

Supports

Metal sulfides are "supported" on materials with high surface areas. A typical support for HDS catalyst is γ-alumina. The support allows the more expensive catalyst to be more widely distributed, giving rise to a larger fraction of the MoS2 that is catalytically active. The interaction between the support and the catalyst is an area of intense interest, since the support is often not fully inert but participates in the catalysis.

Other Uses

The basic hydrogenolysis reaction has a number of uses other than hydrodesulfurization.

Hydrodenitrogenation

The hydrogenolysis reaction is also used to reduce the nitrogen content of a petroleum stream in a process referred to as hydrodenitrogenation (HDN). The process flow is the same as that for an HDS unit.

Using pyridine (C5H5N), a nitrogen compound present in some petroleum fractionation products, as an example, the hydrodenitrogenation reaction has been postulated as occurring in three steps:

Pyridine + Hydrogen	→	Piperdine + Hydrogen	→	Amylamine + Hydrogen	→	Pentane + Ammonia
$C_5H_5N + 5H_2$	→	$C_5H_{11}N + 2H_2$	→	$C_5H_{11}NH_2 + H_2$	→	$C_5H_{12} + NH_3$

and the overall reaction may be simply expressed as:

Pyridine + Hydrogen	→	Pentane + Ammonia
$C_5H_5N + 5H_2$	→	$C_5H_{12} + NH_3$

Many HDS units for desulfurizing naphthas within petroleum refineries are actually simultaneously denitrogenating to some extent as well.

Saturation of Olefins

The hydrogenolysis reaction may also be used to saturate or convert olefins (alkenes) into paraffins (alkanes). The process used is the same as for an HDS unit.

As an example, the saturation of the olefin pentene can be simply expressed as:

Pentene + Hydrogen	→	Pentane
$C_5H_{10} + H_2$	→	C_5H_{12}

Some hydrogenolysis units within a petroleum refinery or a petrochemical plant may be used solely for the saturation of olefins or they may be used for simultaneously desulfurizing as well as denitrogenating and saturating olefins to some extent.

Hydrogenation in the Food Industry

The food industry uses hydrogenation to completely or partially saturate the unsaturatedfatty acids in liquid vegetable fats and oils to convert them into solid or semi-solid fats, such as those in margarine and shortening.

References

- Milton R. Beychok (2005). Fundamentals of Stack Gas Dispersion (Fourth ed.). self-published. ISBN 0-9644588-0-2
- A. Kayode Coker (2007). Ludwig's Applied Process Design for Chemical And Petrochemical Plants, Volume 1 (4th ed.). Gulf Professional Publishing. pp. 732–737. ISBN 0-7506-7766-X
- Wu, Ying; Carroll, John J. (5 July 2011). Carbon Dioxide Sequestration and Related Technologies. John Wiley & Sons. pp. 128–131. ISBN 978-0-470-93876-8
- James H. Gary and Glenn E. Handwerk (2001). Petroleum Refining: Technology and Economics (4th ed.). CRC Press. ISBN 0-8247-0482-7
- M. S. Vassiliou (2 March 2009). Historical Dictionary of the Petroleum Industry. Scarecrow Press. pp. 459–. ISBN 978-0-8108-6288-3
- Wen-Ching Yang (2003). Handbook of Fluidization and Fluid Particle Systems. CRC Press. ISBN 0-8247-0259-X

- Alex C. Hoffmann; Lewis E. Stein (2002). Gas Cyclones and Swirl Tubes:Principles, Design and Operation (1st ed.). Springer. ISBN 3-540-43326-0
- James. G. Speight (2006). The Chemistry and Technology of Petroleum (4th ed.). CRC Press. ISBN 0-8493-9067-2
- David S.J. Jones and Peter P. Pujado (Editors) (2006). Handbook of Petroleum Processing (First ed.). Springer. ISBN 1-4020-2819-9

Permissions

All chapters in this book are published with permission under the Creative Commons Attribution Share Alike License or equivalent. Every chapter published in this book has been scrutinized by our experts. Their significance has been extensively debated. The topics covered herein carry significant information for a comprehensive understanding. They may even be implemented as practical applications or may be referred to as a beginning point for further studies.

We would like to thank the editorial team for lending their expertise to make the book truly unique. They have played a crucial role in the development of this book. Without their invaluable contributions this book wouldn't have been possible. They have made vital efforts to compile up to date information on the varied aspects of this subject to make this book a valuable addition to the collection of many professionals and students.

This book was conceptualized with the vision of imparting up-to-date and integrated information in this field. To ensure the same, a matchless editorial board was set up. Every individual on the board went through rigorous rounds of assessment to prove their worth. After which they invested a large part of their time researching and compiling the most relevant data for our readers.

The editorial board has been involved in producing this book since its inception. They have spent rigorous hours researching and exploring the diverse topics which have resulted in the successful publishing of this book. They have passed on their knowledge of decades through this book. To expedite this challenging task, the publisher supported the team at every step. A small team of assistant editors was also appointed to further simplify the editing procedure and attain best results for the readers.

Apart from the editorial board, the designing team has also invested a significant amount of their time in understanding the subject and creating the most relevant covers. They scrutinized every image to scout for the most suitable representation of the subject and create an appropriate cover for the book.

The publishing team has been an ardent support to the editorial, designing and production team. Their endless efforts to recruit the best for this project, has resulted in the accomplishment of this book. They are a veteran in the field of academics and their pool of knowledge is as vast as their experience in printing. Their expertise and guidance has proved useful at every step. Their uncompromising quality standards have made this book an exceptional effort. Their encouragement from time to time has been an inspiration for everyone.

The publisher and the editorial board hope that this book will prove to be a valuable piece of knowledge for students, practitioners and scholars across the globe.

Index

www.ingramcontent.com/pod-product-compliance
Lightning Source LLC
LaVergne TN
LVHW080904210926
848485LV00004B/591